Lecture Notes in Physics

Edited by J. Ehlers, München, K. Hepp, Zürich,
H. A. Weidenmüller, Heidelberg, and J. Zittartz, Köln
Managing Editor: W. Beiglböck, Heidelberg

49

William G. Harter
Christopher W. Patterson

A Unitary Calculus for Electronic Orbitals

Springer-Verlag
Berlin · Heidelberg · New York 1976

Authors
Dr. William G. Harter
Dr. Christopher W. Patterson
Instituto de Fisica
Departemento de Eletronica Quantica
Universidade Estadual de Campinas
Caixa Postal 1170
Campinas, S. P./Brazil

Library of Congress Cataloging in Publication Data

Harter, William G 1943-
A unitary calculus for electronic orbitals.

(Lecture notes in physics ; 49)
Bibliography: p.
Includes index.
1. Electrons. 2. Atomic orbitals. 3. Molecular orbitals. 4. Unitary operators. I. Patterson, Chris W., 1946- joint author. II. Title. III. Series.
QC793.5.E624H37 539.7'2112 76-17274

ISBN 3-540-07699-9 Springer-Verlag Berlin · Heidelberg · New York
ISBN 0-387-07699-9 Springer-Verlag New York · Heidelberg · Berlin

Printed in Germany
Printing and binding: Beltz Offsetdruck, Hemsbach/Bergstr.

ACKNOWLEDGEMENT

The authors would like to thank Professor Sergio P.S. Porto for his continuing interest and support which has made this reseach possible.
We would also like to thank Professor Jose Ellis Ripper Filho for the hospitality extended to us by the Institute of Physics.
This work was supported in part by the Centro Tecnico Aeroespacial.

TABLE OF CONTENTS

LIST OF FIGURES

LIST OF TABLES

INTRODUCTION

This will be an attempt to describe in the clearest possible way the structure and application of a new unitary operator calculus[1] for atomic and molecular electronic states. This new operator calculus represents a considerable simplification over the previous Racah calculus used extensively in atomic and nuclear physics to calculate energy levels in multiple electron shells. This simplification is a result of the discovery of formula-algorithms using tableau patterns for evaluating directly a) orbital operators, b) transformations between Slater states and generalized Russel-Saunders states, and c) spin and orbit dependent operators; all of which can be applied to any atomic or molecular configuration. Using this calculus, one avoids the computational labor associated with coefficients of recoupling and parentage or sums over permutations. In the following work we give a complete background and review along with new results which describe common types of shell structure, their mixtures, and spin or orbit dependent operators.

The presentation in Part I includes some review of the important ingredients of the pattern theory, some of which are more or less well known, but which the non-specialist might otherwise find difficult to assemble. This review will include the relevant aspects of the Racah calculus,[2] the principle ideas of the Gelfand basis[3] as given in the pioneering works of Biedenharn,[4,5] Louck,[6] and Moshinsky,[7] and finally the much older permutation theory of Young and Rutherford[8] which first became useful in physics by the work of Weyl,[9,10] Yamanouchi[11] and Jahn[12] and furthur developed in the recent works of Robinson[13] and Goddard.[14] Part of the review will use a simple multi-particle system, the atomic (p) (ℓ=1) sub-shell, to demonstrate explicitly old and new ideas. In this way the pattern bases can be more easily compared term by term with those of other approaches.

In Part II we apply the pattern calculus developed in Part I to complex atomic and molecular configurations. In particular, we treat atomic configurations involving pure and mixed subshells with an analysis of more than half filled shells and the jj-coupling approximation. The application of the unitary basis to molecular configurations in the valence orbital and molecular orbital approximation is then introduced. Finally, we give a general mathematical analysis of unitary tableau bases with discussions of the parentage and shell structure of these states.

It is hoped that the new operator calculus will become available to physicists and quantum chemists who are not necessarily specialists in group theory, but who can nevertheless develop and use it for their own problems. Most of this work has been stimulated by the apparent needs of researchers in modern atomic physics who have expressed concern about the computing labor involved in calculating and storing Racah coefficients and fractional parentage coefficients. Applications of atomic physics to modern problems of plasma diagnostics or laser development, for example, can involve a prohibitive amount of computer work. We hope the following will shorten this considerably without requiring an undue amount of new knowledge or change.

PART I: REVIEW AND BASIC STRUCTURE

1. COMPARISON OF SOME APPROACHES TO THE MULTI-FERMION PROBLEM

The principle mathematical problem at hand is the calculation of operators like coulomb interaction, 2^m-moment fields, spin-orbit and so forth for multi-fermion states. We shall ultimately compare the pattern calculus approach with two other approaches. The physical problems that would involve such mathematics include, (a) the electronic structure of free or nearly free atoms, (b) the electronic structure of molecules or complexes, and (c) the structure of nuclei.

Actually there have been many more than two or three calculational schemes devised, particularilly for studying the structure of molecules and nuclei. So our comparisons are by no means exhaustive and will only serve to put the new approach in some kind of context. Nevertheless, there is one sub-problem that all schemes must face in the beginning; namely that of <u>labeling a complete basis of multi-fermion states that satisfy the Pauli Exclusion Principle</u>. We now compare how three approaches deal with this sub-problem. For the time being we shall focus our attention on the problems involving nearly-free atoms only.

A. THE DIRECT APPROACH: SLATER DETERMINANT BASIS. Here the exclusion principle is satisfied immediately by producing at first anti-symmetric products of one-particle states of spin and orbit. (Eq.1) The labeling is done completely by the individual spin and orbital momentum operators of each electron.

$$\Psi_{\substack{\alpha \\ \alpha' \\ \vdots}} = \begin{vmatrix} \chi_\alpha(r_1) & \chi_\alpha(r_2)\dots \\ \chi_{\alpha'}(r_1) & \chi_{\alpha'}(r_2)\dots \\ \vdots & \vdots \end{vmatrix}, \alpha \equiv \{n, \ell, m, m_s\} \qquad (1)$$

However these base states are not eigenstates of total spin, and neither are they eigenstates of total orbital momentum. The two approaches ((B) and (C)) discussed next use bases that are each special combinations of Slater bases, such that they are all eigenstates of total spin, and have orbital parts that can be treated separately. In the unitary approach (C) one obtains states of definite S merely by using a graphical notation and a system of graphical formulas for matrices that goes with it. Eigenstates of orbital momentum L, or of molecular point symmetry are subsequently made using these formulas. In the Racah approach (B) certain coupling coefficients produce states of definite S and L. However certain difficulties lie with the latter as will be discussed below, after which the unitary approach will be introduced.

B. THE RACAH APPROACH: GROUP THEORY OF ROTATIONS. The group theoretical coefficients used in the Racah approach are reviewed below, along with some accompanying theory. The notation in Eq. 2 - 3 will be used throughout. The coefficients are listed below:

(1) Clebsch Gordan $C^{a\ b\ c}_{\alpha\ \beta\ \gamma}$ or Wigner (3-j) Coefficients $\begin{pmatrix} a & b & c \\ \alpha & \beta & \gamma \end{pmatrix}$ couple two angular momentum states to make a third as in Eq.2, or give the relative overlap of a final state with an operator acting on an initial state as in Eq.3 (Eq.3 is called the Wigner

$$\left|(ab)^c_\gamma\right\rangle = \sum_{\alpha\beta} C^{a\ b\ c}_{\alpha\ \beta\ \gamma} \left|{}^a_\alpha\right\rangle \left|{}^b_\beta\right\rangle = \sum_{\alpha\beta} (-1)^{a-b+\gamma} \sqrt{2c+1} \begin{pmatrix} a & b & c \\ \alpha & \beta & -\gamma \end{pmatrix} \left|{}^a_\alpha\right\rangle \left|{}^b_\beta\right\rangle \qquad (2)$$

$$\left\langle {}^c_\gamma \right| T^a_\alpha \left| {}^b_\beta \right\rangle = C^{a\ b\ c}_{\alpha\ \beta\ \gamma} \langle c \| a \| b \rangle \qquad (3)$$

Eckhart Theorem and the constant $\langle c\|a\|b\rangle$ is known as the "reduced matrix element.")

(2) Racah 6-j $\begin{Bmatrix} a & b & \ell' \\ c & d & \ell \end{Bmatrix}$ Coefficients relate one way to couple three momentums to other ways through Eq. 4.

$$\left|(a(bc)^\ell)^d_\delta\right\rangle = \sum_{\ell'} \begin{Bmatrix} a & b & \ell' \\ c & d & \ell \end{Bmatrix} \left|((ab)^{\ell'} c)^d_\delta\right\rangle \qquad (4)$$

(Since coupling can be done for only a pair at a time, there is more than one way to associate three or more momenta, and this recoupling facility is therefore necessary.)

(3) Fractional Parentage Coefficients $(\ell^n{}_\gamma{}^S L \| \} \ell^{n-1}{}_{\gamma'}{}^{S'}{}_{L'})$ give the relative amounts of "parent" (n-1)-electron states in product with the n-th electron states that will be found in a given (n)-electron state which obeys the Pauli exclusion principle, as shown in Eq.5.

$$|\ell^n{}_\gamma{}^S L\, M\, M_S\rangle = \sum_{L'M'S'M_S'\gamma'} (\ell^n{}_\gamma{}^S L \| \} \ell^{n-1}{}_{\gamma'}{}^{S'}{}_{L'})\; C^{L'\ell L}_{M'mM}\; C^{S'\frac{1}{2}S}_{M_S' s M_S} |\ell^{n-1}{}_{\gamma'}{}^{S'}L' M' M_S'\rangle |\ell m s\rangle \tag{5}$$

The γ's represent Racah's state labeling which will be described briefly below.

Now all these coefficients can become involved in operator calculations involving more than three electrons. These expressions are quite complicated even for a pure ℓ-shell 1-body operator, as shown in Eq. 6a. An example in the (p) sub-shell is displayed in Eq. 6b, and will be found again using the new calculus in Sec. 2.

$$\langle \ell^n{}_{\gamma'}{}^S L' M' | V^k_q | \ell^n{}_\gamma{}^S L\, M\rangle = C^{kLL'}_{qMM'} \cdot$$
$$\left[n \sum_{\gamma'' L'' S''} (\ell^{n-1}{}_{\gamma''}{}^{S''}{}_{L''} \| \} \ell^n{}_\gamma{}^S L)(\ell^{n-1}{}_{\gamma''}{}^{S''}{}_{L''} \| \} \ell^n{}_{\gamma'}{}^{S'}{}_{L'})(-1)^{k+\ell+L''+L'} (2\ell+1)(2L+1) \begin{Bmatrix} \ell & k & \ell \\ L & L'' & L' \end{Bmatrix} \langle \ell \| k \| \ell \rangle \right] \quad \text{where: } \langle \ell \| k \| \ell\rangle = (-1)^k \sqrt{\frac{2k+1}{2\ell+1}} \tag{6a}$$

$$\langle p^3\ {}^2P1 | V^2_0 | p^3\ {}^2D1\rangle = C^{221}_{011} \cdot$$
$$3\left[(p^2D \| \} p^3D)(p^2D \| \} p^3P)\sqrt{15}\begin{Bmatrix} 1 & 2 & 1 \\ 2 & 2 & 1 \end{Bmatrix} - (p^2P \| \} p^3D)(p^2P \| \} p^3P)\sqrt{15}\begin{Bmatrix} 1 & 2 & 1 \\ 2 & 1 & 1 \end{Bmatrix} \right] \langle 1 \| 2 \| 1\rangle$$
$$= -\sqrt{\frac{3}{2}} \tag{6b}$$

There exist formulas and tables for the 3-j and 6-j coefficients[15] and for the fractional parentage coefficients there are tables [16] but no formulas. Having no formulas is quite a drawback even when modern computers are used, since large memories need greater access times. Expressions for spin-orbit or 2-body operators require even more use of a memory.

There are more general objections to this sort of approach that have been echoed from time to time concerning "group theory" since its invention when it was called "Gruppenpest." It is possible that this dislike arose from

the frustration with a supposedly powerful mathematics that seemed to be little more than a namer of states and a keeper of tables.

The approach we are presenting uses group theory, too; in fact much of it is very similar to the Racah and Wigner approach, but the new idea is to show how a powerful notation is simultaneously a perfect labeling system and diagramatic computer of operators and properties of states. To show this new approach in some depth, it is instructive to continue with a review of the Racah labeling so similarities and analogies between the two may be exhibited.

The Racah labeling begins with the assemblage of a complete set of 1-body operators. For the orbit (ℓ) there are $2\ell+1$ 1-body states $\left|\begin{smallmatrix}\ell\\m\end{smallmatrix}\right\rangle$, and hence $(2\ell+1)^2$ independent operators. One obtains exactly this number of operators in matrix form through the definition in Eq. 7.

$$\left\langle {\ell \atop m} \middle| v^k_q \middle| {\ell \atop m'} \right\rangle = (-1)^{\ell-m}\sqrt{2k+1}\begin{pmatrix} \ell & k & \ell \\ -m & q & m' \end{pmatrix}$$

$$= C^{k\ell\ell}_{qm'm}\,(-1)^k\left(\frac{2k+1}{2\ell+1}\right)^{\frac{1}{2}} \qquad (7.)$$

These matrices are listed for $\ell=\frac{1}{2}$ and 1 in Table I, and for $\ell=\frac{1}{2}$ through 4 in Tables II and III. It is convenient to have those matrices for any study where angular momentum is important, since all multiple tensors T^k_q or density operators ρ^k_q differ from v^k_q only by the factor called the "reduced matrix element" which we set equal to $\left[(2k+1)/(2\ell+1)\right]^{\frac{1}{2}}(-1)^k$ according to convention. (There often appear as many conventions as authors.)

Now Racah labeling of many particle states is done through special combinations of the operators $V^k \cdot V^k$ defined in Eq. 8.

$$(V^k \cdot V^k) = \sum_q (-1)^q\, V^k_{-q}\, V^k_q \qquad (8a.)$$

$$V^k_q = \sum_{\alpha=1}^{n} v^k_q \text{ (electron } \alpha) \qquad (8b.)$$

One such operator is $V^1 \cdot V^1$ which is proportional to the square $L \cdot L$ of orbital angular momentum and defines the orbital rotation group (R_3) quantum number L.

Some other operators that Racah used were the "G_2 operator"(for f-shell only), the "pairing operator" P, and the (majorana) "exchange operator " M defined in Eqs. 9.

$$M = \sum_{k=0}^{2\ell} V^k \cdot V^k \qquad (9a.)$$

$$P = \sum_{k=0}^{2\ell} (-1)^k \, V^k \cdot V^k \qquad (9b.)$$

$$G = V^1 \cdot V^1 + V^5 \cdot V^5 \qquad (9c.)$$

These are used to distinguish the different states having the same orbital momentum L as described below.

First the exchange operator M defines the permutational symmetry and total spin S associated with a given Racah orbital state. This is denoted by the famous Young frames which will be explained extensively later, since they are a central ingredient of the unitary calculus. For the time being we will simply quote the rules for these diagrams, and give an example. Each n-electron state is associated with an n-box diagram for its orbital part, and another n-box diagram for its spin part. The spin diagram (Fig. 1 b) consists of no more than two rows of boxes, in which each vertical pair of boxes, wherever the rows overlap, corresponds to a pair of spins coupled to zero; and the remainder of boxes in the single row tell what total spin is represented. (Each single box counts $\frac{1}{2}$.) The orbital diagram (Fig. 1 a) is the conjugate of the spin diagram, i.e. rows are exchanged for columns.

Secondly the pairing operator P serves to distinguish some independent states with the same L and S by isolating states according to definite amounts of "orbital pairing" or "seniority." However the physical significance of P and the curious G is somewhat obscure, and we must look instead at the mathematical labeling task for which they were invented.

Racah set out to label all the ℓ^n states with representations of a chain of groups: $U_{2\ell+1} \supset R_{2\ell+1} \supset \ldots \supset R_3 \supset R_2$. The labeling operators from each group, (These are called invariant operators: $v_0^1 \sim L_z$ for R_2, $(V^1 \cdot V^1) \sim L \cdot L$ for R_3, ... P for $R_{2\ell+1}$, and M for $U_{2\ell+1}$), were meant to be a set of mutually commuting operators whose quantum numbers M, L,..., $[\xi]$, $[\lambda]$ were group representation labels and hence invariant state labels.

Unfortunately <u>there is no such chain of groups</u> from $U_{2\ell+1}$ to the spatial rotations $R_3 \supset R_2$ <u>which will unfailingly label all ℓ^n states with its representations</u>. Indeed Racah's labeling is incomplete for the f-sub shell. So it should not be surprising that mathematical calculations with this basis are troublesome.

The next approach to be described contains a labeling scheme which is "perfect": it gives one just enough labels, no more no less than are needed for a given physical situation.

C. THE UNITARY PATTERN APPROACH

The spirit of the newer approach is to use some of the best ideas of the Racah-Wigner approach while avoiding most of its difficulties. The newer approach begins, as did the older one, by assembling a complete set of 1-body operators. For an ℓ-shell, for example, we define $(2\ell+1)^2$ ELEMENTARY OPERATORS e_{ij}. Each of these is defined by an $(2\ell+1)(2\ell+1)$ matrix that is all zero except for a single (1) at the i,j position. (See Eq. 10)

$$\langle {}^{\ell}_{m'} | e_{ij} | {}^{\ell}_{m} \rangle = \delta_{im'} \ \delta_{jm} \tag{10}$$

Expressing e_{ij} in terms of v^k_q and visa-versa is easily done by inspection of the appropriate entry in Table I. We will need these relations shortly. The same relations hold between the Racah V^k_q (recall Eq. 8b) and the E_{ij} defined by Eq. 11.

$$E_{ij} = \sum_{\alpha=1}^{n} e_{ij} \ (\text{electron } \alpha) \tag{11}$$

These operators are called the generators of the unitary group $U_{2\ell+1}$. They satisfy communication relations of Eq. 12 which follow immediately from the basic definition of the e_{ij}'s. (Eq. 10)

$$[E_{ij}, E_{k\ell}] = \delta_{jk} E_{i\ell} - \delta_{\ell i} E_{kj} \tag{12}$$

These relations are a generalization of relations between angular momentum operators i.e. the generators of the rotation group R_3. In fact the relation $[L_x, L_y] = i L_z$ is a special case of Eq. 12 if you use Eq. 13 which follows partly from the first table Ia.

$$\begin{aligned} L_x &\equiv (V^1_{-1} - V^1_1) / 2 = (E_{\frac{1}{2}\frac{-1}{2}} + E_{\frac{-1}{2}\frac{1}{2}}) / 2 \\ L_y &\equiv (-V^1_{-1} - V^1_1) / 2 = (E_{\frac{1}{2}\frac{-1}{2}} - E_{\frac{-1}{2}\frac{1}{2}}) / 2i \\ L_z &\equiv V^1_0 / \sqrt{2} = (E_{\frac{1}{2}\frac{1}{2}} - E_{\frac{-1}{2}\frac{-1}{2}}) / 2 \end{aligned} \tag{13}$$

All of the Racah Wigner angular momentum calculus follows from the relations such as $[L_x, L_y] = iL_z$, and the beautiful generalization of this amounts to a "giant" calculus involving all operators E_{ij} (or V^k_q) over many-particle states which follows analogously from the relations in Eq. 12. However this generalization can be made only when the old problem of labeling states is solved, so we describe this labeling now.

A representation of U_m is labeled by m integers $\lambda_{1,m}\ \lambda_{2,m}\ \lambda_{3,m} \ldots \lambda_{m,m}$ which give the maximum "quantum numbers" that can be acheived in a single state by the eigenvalues of E_{11}, E_{22}, ... , and E_{mm} , subject to an ordering convention $\lambda_{1,m} \geqslant \lambda_{2,m} \geqslant \ldots\ldots \geqslant \lambda_{m,m}$. (Analogously, the representation label L of the rotation group R_3 gives the highest eigenvalue M of L_z.) Next, the representations of U_{m-1} contained in $(\lambda_{1,m}\ \lambda_{2,m}\ \lambda_{3,m} \ldots \lambda_{m,m})$ are labeled by (m-1) integers $\lambda_{1,m-1}\ \lambda_{2,m-1} \ldots \lambda_{m-1,m-1}$ where these $\lambda_{i,m-1}$ vary while subject to "betweeness conditions" of Eq. 14.

$$\lambda_{1,m} \geqslant \lambda_{1,m-1} \geqslant \lambda_{2,m} \quad \lambda_{2,m} \geqslant \lambda_{2,m-1} \geqslant \lambda_{3,m} \ldots \\ \ldots \lambda_{m-1,m} \geqslant \lambda_{m-1,m-1} \geqslant \lambda_{m,m} \tag{14}$$

(To carry our analogy furthur, magnetic quantum number M labels various representations of cylindrical rotation symmetry R_2 that are contained in the L representation of R_3. M varies over integral steps but stays "between" L and -L.)

Now labels of the representations of $U_m \supset U_{m-1} \supset U_{m-2} \cdots \supset U_1$ are assembled into an m-row triangular pattern which is called the Gelfand Pattern (Fig. 2a) and this constitutes a complete labeling of all U_m states. However, for most physical applications, particularily in atomic and molecular theories, this is more than enough labels, and the unused ones would just get in the way.

Another completely equivalent notation involving Young Tableaus avoids this redundancy by letting the λ_{ij} integers be represented as lengths of rows of boxes containing certain numbers as shown in Fig. 2b. Now when a label is redundant, it simply vanishes from sight. This is the first of many conveniences of the box pattern notation which will be described in the next section.

Before continuing, we will exhibit one example of the simple mechanics of this labeling. Take the representation (210) of U_3 which will be considered later. Upon breaking the symmetry to U_2 we find, according to Eq. 14, that the representations (21), (20), (11), and (10) would appear on the diagonal of the (210) matrix schematicized in Fig. 3. Then reduction to U_1 would give (2) and (1) inside (21); (2), (1), and (0) inside (20), etc. for a total of eight 1-dimensional "eigenvalues" in all.

It is in this way that the representation (210) labels its own eight bases which are pictured à la Gelfand and Young alongside the matrix in Fig. 3. It should be clear how pattern labeling in general assures all states a distinct genealogy, i.e. labeling. It should also be clear that the Gelfand "betweeness" condition (Eq. 14) forces all the

Young tableaus to be "lexical" i.e. have numbers ordered to increase only toward the right in the rows and down in the columns of boxes, with no numbers repeated in a single column. A complete set of "lexical" tableau define a complete basis as well as Eq. 14.

2. MECHANICS OF THE UNITARY CALCULUS: EXAMPLES

Inthis section we will try to expose all the mathematical machinery that we know how to apply, in detail, using the simplest non-trivial examples of each case in point.

A. Spin $\frac{1}{2}$: U_2

A tableau description of spin $\frac{1}{2}$ is necessarily a review of some "well known" facts, except for a few points about pattern calculus which we need to bring out in context. First of all, it is true that any unitary combination of state 1 (spin up↑) and state 2 (spin down↓) is just a rotation away from any other state. Indeed the only U(2) generators (See Table I-a) besides the identity are the V^1_q generators of rotation, so U(2) theory is mathematically equivalent to angular momentum calculus. However if more than two spin $\frac{1}{2}$ states are coupled using the Wigner and Racah coefficients a "log-jam" can easily result. Instead we show how the states are organized according to tableaus.

One main idea of the tableau is to exhibit the permutational symmetry of an n-particle state. For example, the two-electron state that is anti-symmetric (S=0) is designated by two boxes stacked vertically (11) (See Fig.4), while the three symmetric states (S=1) are all labeled by two boxes in a row (20). In other words, these Young Tableaus also label representations of the chain of permutation groups $S_1 \subset S_2 \subset S_3$ ... in practically the same way that they label representations of $U_1 \subset U_2 \subset U_3$. In fact they were invented for S_n long before the study of U_m or quantum mechanics began.

Readers unfamiliar with the S_n groups may appreciate knowing that the first few are also well known crystal point groups. S_2, of course, is the same as two-fold symmetry C_2, S_3 is the same as trigonal symmetries D_3 (3:2), and S_4 is tetrahedral symmetry T_d(3/4). (Higher dimensional point symmetries correspond to higher S_n.)[17] The representation bases for these groups are labeled in Fig. 5 by tableaus and more common notations. Using this may help explain how, for example, the two different (210) or "E" symmetry ($S=\frac{1}{2}$) states in Fig. 4 (four states in all) arise from 3 electrons.This is the first example where the permutation representation ($D^{210}(p)$) is two-dimensional, and for each of the corresponding two values of μ in the group projection operator of Eq. 15, there will arise a complete $S=\frac{1}{2}$ doublet.

$$\Psi_{M_S}^{S=\frac{1}{2}}(\mu) = N_\nu \bar{P}^{210}_{\mu\nu}\, \chi^{\frac{1}{2}}_{m_s}(1)\, \chi^{\frac{1}{2}}_{m'_s}(2)\, \chi^{\frac{1}{2}}_{m''_s}(3)$$
$$= N_\nu \frac{2}{3!} \sum_{\substack{\text{permutations} \\ p}} D^{210}_{\mu\nu}(p)\, \bar{p}\, \chi^{\frac{1}{2}}_{m_s}(1)\, \chi^{\frac{1}{2}}_{m'_s}(2)\, \chi^{\frac{1}{2}}_{m''_s}(3) \quad (15)$$

(The ν in Eq. 15 is tied to the U(2) quantum number values $m_s + m'_s + m''_s = M_s$ as explained in Sec. 3 when we investigate the general properties of these projections. The N_ν is a normalization derived there also.)

Now it might appear that we are getting ourselves into the worst "Gruppenpest" of all. For with only 10 particles we would face a group of order 3,628,800 and fill libraries with tables about it. Apparently though, we can be saved all this by the incredible fact that these patterns have practically all the information you want written on their faces.

As a first example of this information system, we review the "hook-formulas" of Robinson[18] and Hall (See Fig.6). These give the dimensions of representations of S_n groups and U_m groups. We see how they apply to finding the number of states in Fig. 4, depending on whether we count the number of spin components (dimension of a U_2 representation) or the number of ways to make a given spin component (dimension of a S_n representation).

It is this type of non-algebraic information storage that interests us. For more and more complex multi-particle shells, any algebraic description is bound to become prohibitively complicated, so it is very worthwhile to uncover algorithms that will moderate this complexity by automatically ignoring what is redundent or zero for each individual case. To date, several examples have been found of algorithms of the same general nature as the Hall - Robinson formulas. These include a character formula by Coleman[19], normalization formulas of Biedenharn and Ciftan[20], and S_n representation formulas of Yamanouchi[21] We describe the Yamanouchi formula now since it is closely related to two algorithms which we have invented for dealing with atomic and molecular operators, and which will be treated next.

The Yamanouchi formula (See Fig. 7) gives the representations of the principle or "nearest-neighbor" permutations (n, n-1). All other permutations are just products of these. The formula depends upon the relative positions of the two boxes that contain n and n-1 respectively in each tableau, i.e. the axial or "city block" distance d between them. This should be familiar to urban dwellers as the minimum number of blocks you must walk or streets you must cross to go from one block center to the other. (Several zig-zag paths may have this same length d.) If the two boxes are in the same row or column, then they must be neighbors (d=1) and a (n, n-1) switch gives no legal ("lexical") tableau. But, the permutation has a diagonal matrix element of

-1 or +1 depending on whether n-1 is above or to the left of n, as seen in Fig. 7.

If the permutation (n,n-1) does give a lexical tableau, then the diagonal elements of the representation are -1/d or +1/d depending upon whether n-1 is above or to the left of n, and there will be the off-diagonal components shown in Fig.7.

B. Orbit $\ell=1$; U_3

Now an important part of the pattern calculus is demonstrated using the atomic p sub-shell as an example. The three single-particle orbital states : $|1\rangle = \left|{\ell=1 \atop m=1}\right\rangle$, $|2\rangle = \left|{\ell=1 \atop m=0}\right\rangle$, and $|3\rangle = \left|{\ell=1 \atop m=-1}\right\rangle$ span the fundamental or defining representation of U_3. Now we find states like $|2\rangle$ or$(|1\rangle + |3\rangle)/\sqrt{2}$ of "plane" polarization that are not just rotations of all other states such as $|1\rangle$ of "circular" polarization. Now we need other operators besides the generators v^1_q of rotation that "did everything" for a spin $\frac{1}{2}$ particle. This complete set for $\ell=1$ (See Table Ib) includes the five quadrupole operators v^2_q and generates the group U_3.

The orbital states of the p sub-shell are displayed in Fig. 8 by a diagram involving the lexical U_3 tableaus. Each state is plotted as a point in a three dimensional coordinate system in which the 1-component of a state is the number of 1's in the tableau and the same for the 2's and 3's.

In order to find matrix elements of the physical operators V^k_q , we must identify them with the elementary operators E_{ij} , using Table Ib. First the V^k_0 operators are always represented by diagonal matrices, since they are linear combinations of E_{11} , E_{22} ,and E_{33} only, whose eigenvalues are just the number of 1's, 2's, and 3's, respectively, of each state plotted in Fig. 8. In fact there is a (V_0^0, V_0^1, V_0^2) coordinate system defined by Table Ib or Eq. 16 below, and two of its axes are shown in Fig. 8.

$$\sqrt{3}\, V_0^0 = E_{11} + E_{22} + E_{33} \tag{16 a}$$
$$\sqrt{2}\, V_0^1 = E_{11} - E_{33} \equiv L_z \tag{16 b}$$
$$\sqrt{6}\, V_0^2 = E_{11} - 2E_{22} + E_{33} \tag{16 c}$$

The other operators V_q^k for $q \neq 0$ correspond to the "raising and lowering" operators E_{ij} for $i \neq j$, as given in Table Ib or Eq. 17 below.

$$V_2^2 = E_{13} \tag{17 a}$$
$$-2V_1^2 = \sqrt{2}(E_{12} - E_{23}) \tag{17 b}$$
$$-2V_1^1 = \sqrt{2}(E_{12} + E_{23}) \equiv L_+ \tag{17 c}$$
$$2V_{-1}^1 = \sqrt{2}(E_{21} + E_{32}) \equiv L_- \tag{17 d}$$
$$2V_{-1}^2 = \sqrt{2}(E_{21} - E_{32}) \tag{17 e}$$
$$V_{-2}^2 = E_{31} \tag{17 f}$$

It is helpful to think of E_{ij} as an operator that "destroys" a j and "creates" an i, while E_{jj} is a number operator that counts the number of j's.

Matrix elements of E_{ij} exist only between lexical tableaus that differ solely in the replacement of a j by an i.

These matrix elements are found using the simple orbital pattern formulas of Fig. 9. The "principle elementary operators" $E_{i-1,i}$ are given there (Changing a 2 to a 1, i.e. m=0 to m=1, is called "raising" ...) and all others are products of these, according to Eq. 12 and Eq. 18, or transposes, according to Fig. 9b.

$$\begin{aligned} E_{i-2,i} &= \left[E_{i-2,i-1}, E_{i-1,i}\right] \\ E_{i-3,i} &= \left[E_{i-3,i-2}, E_{i-2,i}\right] \\ &\vdots \end{aligned} \tag{18}$$

The conventional angular momentum raising operator L_+ (Eq. 17c), lowering operator L_- (Eq. 17d) and z-component (Eq. 16b) lead to the conventional ${}^{2S+1}L$ identification of the tableaus. The numbers in each tableau give its M-eigenvalue (of L_z) or its $2V_0^1$ component in Fig.8, and it is thereby seen which ${}^{2S+1}L$ terms are allowed for a particular multiplet.

Consider the $(p)^3$ configuration (Nitrogen) in Fig.8. The highest (spin) doublet, or (210) tableau, has M=2 so it must be a ^2D as in Eq.19, since it is alone.

$$\left|{}^2D\ 2\right\rangle = \begin{matrix} 1\,1 \\ 2 \end{matrix} \tag{19}$$

The two elementary components of L_- (Eq. 17 d) are E_{21} and E_{32}. Sequential application of these gives all the tableau states in the multiplet and shows spectral terms allowed. Application of L_- according to Eq. 20 leads to the correct state vector for each of the terms.

$$\frac{L_- \left| {}^{2S+1}L\ M \right\rangle}{\sqrt{(L+M)(L-M+1)}} = \left| {}^{2S+1}L\ M-1 \right\rangle \qquad (20)$$

For example E_{21} and E_{32} on state $\begin{smallmatrix}11\\2\end{smallmatrix}$ gives $\begin{smallmatrix}12\\2\end{smallmatrix}$ and $\begin{smallmatrix}11\\3\end{smallmatrix}$ respectively, both with matrix elements of unity according to Fig. 9-h and 9-g respectively. Then operation with L_- gives Eq. 21 below.

$$\left|{}^2D\ 1\right\rangle = \frac{L_- \left|{}^2D\ 2\right\rangle}{2} = \left(\begin{smallmatrix}12\\2\end{smallmatrix} + \begin{smallmatrix}11\\3\end{smallmatrix} \right)/\sqrt{2} \qquad (21)$$

Since there are two M=1 tableaus, there must also be an L=1 state, namely the orthogonal ^2P state given in Eq. 22 below.

$$\left|{}^2P\ 1\right\rangle = \left(\begin{smallmatrix}12\\2\end{smallmatrix} - \begin{smallmatrix}11\\3\end{smallmatrix} \right)/\sqrt{2} \qquad (22)$$

Now the quadrupole (V^2_0) matrix element given by the Racah expression in Eq. 6 b is computed using Fig. 9 a in Eq. 23 below. Note in contrast that one gets all k-moments from one expression involving a few modified Wigner coefficients $({}_i{}^k{}_j)$. This coefficient is the ij th component of the v^k that has this component in Table Ib.

$$\left\langle {}^2P\ 1 \right| v^k_0 \left| {}^2D\ 1 \right\rangle = \frac{\begin{smallmatrix}12\\2\end{smallmatrix} - \begin{smallmatrix}11\\3\end{smallmatrix}}{\sqrt{2}} \left(({}_1{}^k{}_1)E_{11} + ({}_2{}^k{}_2)E_{22} + ({}_3{}^k{}_3)E_{33} \right) \frac{\begin{smallmatrix}12\\2\end{smallmatrix} + \begin{smallmatrix}11\\3\end{smallmatrix}}{\sqrt{2}}$$

$$\begin{aligned} &= \tfrac{1}{2}\left(-({}_1{}^k{}_1)+2({}_2{}^k{}_2)-({}_3{}^k{}_3)\right) = -\sqrt{\tfrac{3}{2}} \quad \text{for } k=2 \\ &= 0 \quad k=1 \\ &= 0 \quad k=0 \end{aligned} \qquad (23)$$

Eq. 24 gives all the matrix elements for the remaining $S=\frac{1}{2}$ or doublet states. The superscript $^{(ij)}$ above each matrix element in this equation indicates which operator E_{ij} was responsible. (No more than one can occupy a position with $i \neq j$.)

	11 2 (M=2)	12 2 (M=1)	11 3	12 3 (M=0)	13 2	13 3	22 3	23 3
11 2	$\overset{(11)\,(22)}{2+1}$	$1^{(12)}$	$1^{(23)}$	$-\sqrt{\frac{1}{2}}^{(13)}$	$\sqrt{\frac{3}{2}}^{(13)}$	.	.	.
12 2	.	$\overset{(11)\,(22)}{1+2}$	.	$\sqrt{\frac{1}{2}}^{(23)}$	$\sqrt{\frac{3}{2}}^{(23)}$	.	$-1^{(13)}$	.
11 3	.	.	$\overset{(11)\,(33)}{2+1}$	$\sqrt{2}^{(12)}$	.	$1^{(13)}$	.	.
12 3	.	.	.	$\overset{(11)\,(22)\,(33)}{1+2+3}$	.	$\sqrt{\frac{1}{2}}^{(23)}$	$\sqrt{2}^{(12)}$	$\sqrt{\frac{1}{2}}^{(13)}$
13 2	.	.	.	.	$\overset{(11)\,(22)\,(33)}{1+2+3}$	$\sqrt{\frac{3}{2}}^{(23)}$	.	$\frac{\sqrt{3}}{\sqrt{2}}^{(13)}$
13 3	.	.	.	.	.	$\overset{(11)\,(33)}{1+2}$	.	$1^{(12)}$
22 3	.	.	.	.	.	.	$\overset{(22)\,(33)}{2+1}$	$1^{(23)}$
23 3	.	.	.	.	.	.	.	$\overset{(22)\,(33)}{1+2}$

$= \langle E_{ij} \rangle$ for $i<j = 1,2,\text{or}\,3$ (24)

For the quartet terms $(S=\frac{3}{2})$ there is only one allowed orbital tableau, (Eq. 25) and for it only the V_0^0 matrix is non-zero.

$$|{}^4S\ 0\rangle = \begin{matrix}1\\2\\3\end{matrix} \tag{25}$$

These procedures will be generalized slightly in Part II in order to account for repeated states of the same L and S. (However no labeling problem needs to be solved ever again since the tableaus are quite complete in this respect.)

Isotropic two-body operators such as electrostatic repulsion are combinations of operator products $E_{ij}E_{ji}$ contained in the familiar scalar products $V^k \cdot V^k$. (This will be reviewed in detail for the general problem of multiple shells in Part II.)

It turns out that the matrix elements of the scalars can be evaluated by inspection of an E-matrix like Eq. 24. The matrix element of $V^k \cdot V^k$ between tableau state γ and state γ' is equal to the overlap or scalar product of column γ with column γ' in Eq.24 plus a similar overlap of row γ with row γ'. (The diagonal is counted just once for the case $\gamma=\gamma'$, however.) For example the components needed to compute the electrostatic energy e (In Eq. 26 the constants r and r' are radial integrals.) for the p^3 configuration are given in Eq. 27 and 28.

$$e = r(V^2 \cdot V^2) + r' \qquad (26)$$

Note that the superscripts (ij) in Eq. 24 are transcribed into Eq. 27 as the coefficients $({}_i{}^k{}_j)$ found in table I-b.

$$\begin{aligned}
\left\langle {}^{11}_{2} \right| V^k \cdot V^k \left| {}^{11}_{2} \right\rangle &= (2({}_1{}^k{}_1)+({}_2{}^k{}_2))^2+({}_1{}^k{}_2)^2+({}_2{}^k{}_3)^2+2({}_1{}^k{}_3)^2 \\
\left\langle {}^{12}_{2} \right| V^k \cdot V^k \left| {}^{12}_{2} \right\rangle &= (({}_1{}^k{}_1)+2({}_2{}^k{}_2))^2+({}_1{}^k{}_2)^2+2({}_2{}^k{}_3)^2+({}_1{}^k{}_3)^2 \\
\left\langle {}^{12}_{2} \right| V^k \cdot V^k \left| {}^{11}_{3} \right\rangle &= 2({}_1{}^k{}_2)({}_2{}^k{}_3) \qquad (27) \\
\left\langle {}^{11}_{3} \right| V^k \cdot V^k \left| {}^{11}_{3} \right\rangle &= (2({}_1{}^k{}_1)+({}_3{}^k{}_3))^2+2({}_1{}^k{}_2)^2+({}_2{}^k{}_3)^2+({}_1{}^k{}_3)^2
\end{aligned}$$

$$\langle V^2 \cdot V^2 \rangle = \begin{array}{c|ccc} & {}^{11}_{2} & {}^{12}_{2} & {}^{11}_{3} \\ \hline {}^{11}_{2} & 3 & \cdot & \cdot \\ {}^{12}_{2} & \cdot & 4 & -1 \\ {}^{11}_{3} & \cdot & -1 & 4 \end{array} \qquad (28)$$

The $V^2 \cdot V^2$ eigenvalues of 3 for ^2D, 5 for ^2P (We use Eq.22), and 0 for ground state ^4S, give the first approximate predictions for the spacing of the lowest three levels of Nitrogen.

As will be shown in Part II, orthogonal states with the same L and S give off-diagonal components for the electrostatic operator.

C. Spin $\frac{1}{2}$ and Orbit $\ell=1$ Together : $U_2 \times U_3$

The discussion of the orbital part of the electrons in Sec. 2B has it "disembodied" from the spin part and visa-versa in Sec. 2A. The reincarnated state that obeys the exclusion principle is represented by assembling with each orbital tableau all the spin tableaus of the "conjugate" shape, (Recall Fig.1) as in Eq.29.

$$\left|\begin{array}{ll} 11 & \uparrow\uparrow\uparrow \\ 2 & \downarrow \\ 3 & \end{array}\right\rangle , \left|\begin{array}{ll} 11 & \uparrow\uparrow\downarrow \\ 2 & \downarrow \\ 3 & \end{array}\right\rangle , \left|\begin{array}{ll} 11 & \uparrow\downarrow\downarrow \\ 2 & \downarrow \\ 3 & \end{array}\right\rangle \\ \vdots \\ \left|\begin{array}{ll} 11 & \uparrow\uparrow \\ 22 & \downarrow\downarrow \end{array}\right\rangle \\ \vdots \qquad (29)$$

Symbols like the above represent a generally very complicated but precisely defined multi-electron Pauli state which is described one way in Sec. 2D, and another way in Sec. 3. But it turns out that the complicated sums implicit to this reincarnation can be ignored as long as the physical significance of each part is known. For example, combination of the appropriate tableaus in Sec. 2A-2B yield the Russell-Saunders (LS) states of a (Nitrogen) $(p)^3$ configuration in Eq.30

$$\left|{}^4S\ \begin{array}{ll} L=0 & S=3/2 \\ M=0 & M_s=3/2 \end{array}\right\rangle = \left|\begin{array}{l} 1 \\ 2 \\ 3 \end{array}\ \begin{array}{l}\uparrow\uparrow\uparrow\end{array}\right\rangle \tag{30}$$

$$\left|{}^4S\ \begin{array}{ll} 0 & 3/2 \\ 0 & 1/2 \end{array}\right\rangle = \left|\begin{array}{l} 1 \\ 2 \\ 3 \end{array}\ \begin{array}{l}\uparrow\uparrow\downarrow\end{array}\right\rangle$$

$$\vdots$$

$$\left|{}^2P\ \begin{array}{ll} 1 & 1/2 \\ 1 & 1/2 \end{array}\right\rangle = \left[\left|\begin{array}{ll} 12 & \uparrow\uparrow \\ 2 & \downarrow \end{array}\right\rangle - \left|\begin{array}{ll} 11 & \uparrow\uparrow \\ 3 & \downarrow \end{array}\right\rangle\right]/\sqrt{2}$$

$$\left|{}^2P\ \begin{array}{ll} 1 & 1/2 \\ 1 & -1/2 \end{array}\right\rangle = \left[\left|\begin{array}{ll} 12 & \uparrow\downarrow \\ 2 & \downarrow \end{array}\right\rangle - \left|\begin{array}{ll} 11 & \uparrow\downarrow \\ 3 & \downarrow \end{array}\right\rangle\right]/\sqrt{2}$$

$$\left|{}^2P\ \begin{array}{ll} 1 & 1/2 \\ 0 & 1/2 \end{array}\right\rangle = \left[-\left|\begin{array}{ll} 12 & \uparrow\uparrow \\ 3 & \downarrow \end{array}\right\rangle + \sqrt{3}\left|\begin{array}{ll} 13 & \uparrow\uparrow \\ 2 & \downarrow \end{array}\right\rangle\right]/\ 2$$

$$\vdots$$

$$\left|{}^2D\ \begin{array}{ll} 2 & 1/2 \\ 2 & 1/2 \end{array}\right\rangle = \left|\begin{array}{ll} 11 & \uparrow\uparrow \\ 2 & \downarrow \end{array}\right\rangle$$

$$\vdots$$

The orbital operators work on orbital parts of states (29)-(30) while ignoring the spin parts, and visa-versa. This is part of what is implied by the notation $U_3 \times U_2$; operators (like L) in U_3 commute with those (like S) in U_2.

Now our final problem involves operators like spin-orbit interaction which fail to commute with L or S, and exist outside of $U_3 \times U_2$. A pattern calculus for a bigger group U_6 covers this, as described next.

D. Spin $\frac{1}{2}$ and Orbit $\ell=1$ Together: U_6

A single electron confined to the p-shell has six possible states which we label as follows: $|a\rangle = |1\uparrow\rangle = \left|\begin{array}{ll} \ell=1 & S=\frac{1}{2} \\ m=1 & \sigma=\frac{1}{2} \end{array}\right\rangle$, $|b\rangle = |1\downarrow\rangle$, $|c\rangle = |2\uparrow\rangle$, $|d\rangle = |2\downarrow\rangle$, $|e\rangle = |3\uparrow\rangle$, and $|f\rangle = |3\downarrow\rangle$. Along with this comes a set of 36 1-body operators which could be the U_6 elementary operators $E_{aa}\ E_{ab} \cdots E_{ff}$, or else Racah's double tensor operators $v^{k\lambda}_{q\sigma}$, for which a definition and examples to be discussed are given in Eq.31.

$$\left\langle \begin{matrix} \ell & \frac{1}{2} \\ m' & \mu' \end{matrix} \right| \; v^{k\lambda}_{q\sigma} \; \left| \begin{matrix} \ell & \frac{1}{2} \\ m & \mu \end{matrix} \right\rangle = \left\langle \begin{matrix} \ell \\ m' \end{matrix} \right| \; v^{k}_{q} \; \left| \begin{matrix} \ell \\ m \end{matrix} \right\rangle \left\langle \begin{matrix} \frac{1}{2} \\ \mu' \end{matrix} \right| \begin{matrix} \lambda \\ v_\sigma \end{matrix} \left| \begin{matrix} \frac{1}{2} \\ \mu \end{matrix} \right\rangle \qquad (31a)$$

$$v^{11}_{11} = -\sqrt{\tfrac{1}{2}} \begin{bmatrix} \cdot & \cdot & \cdot & \cdot & \cdot & \cdot \\ \cdot & \cdot & 1 & \cdot & \cdot & \cdot \\ \cdot & \cdot & \cdot & \cdot & \cdot & \cdot \\ \cdot & \cdot & \cdot & \cdot & 1 & \cdot \\ \cdot & \cdot & \cdot & \cdot & \cdot & \cdot \\ \cdot & \cdot & \cdot & \cdot & \cdot & \cdot \end{bmatrix} = -\sqrt{\tfrac{1}{2}}(e_{bc} + e_{de}) \qquad (31b)$$

$$v^{11}_{00} = \tfrac{1}{2} \begin{bmatrix} 1 & \cdot & \cdot & \cdot & \cdot & \cdot \\ \cdot & -1 & \cdot & \cdot & \cdot & \cdot \\ \cdot & \cdot & \cdot & \cdot & \cdot & \cdot \\ \cdot & \cdot & \cdot & \cdot & \cdot & \cdot \\ \cdot & \cdot & \cdot & \cdot & -1 & \cdot \\ \cdot & \cdot & \cdot & \cdot & \cdot & 1 \end{bmatrix} = \tfrac{1}{2}(e_{aa} - e_{bb} - e_{ee} + e_{ff}) \qquad (31c)$$

Note that Eq.31(b-c) are derived from products of tables Ia with Ib, and that operators of $U_3 \times U_2$ are v^{k0}_{q0} and $v^{0\lambda}_{0\sigma}$.

Now the exclusion principle permits us to have only totally anti-symmetric multi-electron U_6 states i.e. states with tableau of one column only. The matrix elements of operator E_{ij} between these Slater Determinant states are 0 if there is no j to be destroyed or an i already present in initial state, -1 if the new i must be moved an odd number of boxes to get "lexical" order, +1 if the new i is within an even substitution of order.

$$E_{cf} \begin{matrix} a \\ c \\ d \\ e \\ f \end{matrix} = 0 \;, \quad E_{bf} \begin{matrix} a \\ c \\ d \\ e \\ f \end{matrix} = - \begin{matrix} a \\ b \\ c \\ d \\ e \end{matrix} \;, \quad E_{be} \begin{matrix} a \\ c \\ d \\ e \\ f \end{matrix} = \begin{matrix} a \\ b \\ c \\ d \\ f \end{matrix} \qquad (32)$$

Now, the tableau formula in Fig.10 gives the coefficients to transform between these vertical tableau

U_6 states and the $U_3 \times U_2$ or L-S states of Eqs. 29 or 30. Correct normalization and sign are automatically given. A good deal of labor can be saved by this formula, in fact there are only a few transformations like Eq. 33 for which the coefficients can quickly be checked by simple matrix arithimetic.

	$12\downarrow\uparrow$	$\begin{matrix}1\\2\end{matrix}\uparrow\downarrow$
$1\uparrow\ 2\downarrow$	$\sqrt{1/2}$	$\sqrt{1/2}$
$1\downarrow\ 2\uparrow$	$-\sqrt{1/2}$	$\sqrt{1/2}$

(33)

The spin-orbit operator of Eq. 34 follows from the relation in Eq. 31.

$$\begin{aligned} \text{S.O.} &= \zeta \sum_{\alpha=1}^{n} \vec{s}(\text{electron }\alpha)\cdot\vec{\ell}(\text{electron }\alpha) \\ &= \zeta\left(V^{1\ 1}_{0\ 0} - V^{1\ 1}_{1 -1} - V^{1\ 1}_{-1\ 1} \right) \\ &= \zeta\left[\tfrac{1}{2}(E_{aa} - E_{bb} - E_{ee} + E_{ff}) + \sqrt{\tfrac{1}{2}}(E_{bc} + E_{de} + E_{cb} + E_{ed})\right] \end{aligned} \quad (34)$$

This operator is invariant to rigid rotations of both spin and orbit, so it is convenient to deal with the LSJ states made from Eq. 30 and Eq. 35.

$$|{}^{2S+1}L_J\ M_J\rangle = \sum_{M,M_S} C^{L\ S\ J}_{M\ M_S M_J}\ |L\ M\rangle|S\ M_S\rangle \quad (35)$$

The maximum M_J examples of the states in Eq. 35 are written explicitly for the $(p)^3$ configuration in Eq. 36.

$$\left|{}^2D_{\frac{5}{2}}\ \tfrac{5}{2}\right\rangle \equiv \begin{array}{c}a\\b\\c\end{array} \qquad \left|{}^2D_{\frac{3}{2}}\ \tfrac{3}{2}\right\rangle = \sqrt{\tfrac{4}{5}}\begin{array}{c}a\\b\\d\end{array} + \sqrt{\tfrac{1}{10}}\left(\begin{array}{c}a\\c\\d\end{array} - \begin{array}{c}a\\b\\e\end{array}\right) \qquad \left|{}^2P_{\frac{1}{2}}\ \tfrac{1}{2}\right\rangle = -\sqrt{\tfrac{1}{3}}\begin{array}{c}b\\c\\d\end{array} - \sqrt{\tfrac{1}{3}}\begin{array}{c}a\\b\\f\end{array} - \sqrt{\tfrac{1}{6}}\begin{array}{c}a\\c\\f\end{array} + \sqrt{\tfrac{1}{6}}\begin{array}{c}b\\c\\e\end{array}$$

$$\left|{}^2P_{\frac{3}{2}}\ \tfrac{3}{2}\right\rangle = -\sqrt{\tfrac{1}{2}}\left(\begin{array}{c}a\\c\\d\end{array} + \begin{array}{c}a\\b\\e\end{array}\right)$$

$$\left|{}^4S_{\frac{3}{2}}\ \tfrac{3}{2}\right\rangle = \begin{array}{c}a\\c\\e\end{array} \tag{36}$$

The matrix elements of the spin-orbit (S.O.) operator in Eq. 34 are now made easily according to the determinant or tableau rules previously described. Examples of their application are shown in Eq. 37.

$$\left\langle {}^2P_{\frac{3}{2}}\right| \text{S.O.} \left| {}^4S_{\frac{3}{2}}\right\rangle = \frac{-\left(\begin{array}{c}a\\c\\d\end{array} + \begin{array}{c}a\\b\\e\end{array}\right)}{\sqrt{2}}(\text{S.O.})\left(\begin{array}{c}a\\c\\e\end{array}\right) = -\zeta\,\frac{\left(\begin{array}{c}a\\c\\d\end{array} + \begin{array}{c}a\\b\\e\end{array}\right)}{\sqrt{2}}\cdot\frac{\left(\begin{array}{c}a\\b\\e\end{array} + \begin{array}{c}a\\c\\d\end{array}\right)}{\sqrt{2}} \tag{37a}$$

$$= -\zeta$$

$$\text{S.O.} = \zeta \begin{array}{c} \begin{array}{ccccc} {}^2D_{\frac{5}{2}} & {}^2D_{\frac{3}{2}} & {}^2P_{\frac{3}{2}} & {}^4S_{\frac{3}{2}} & {}^2P_{\frac{1}{2}} \end{array} \\ \boxed{\begin{array}{ccccc} 0 & \cdot & \cdot & \cdot & \cdot \\ \cdot & 0 & -\sqrt{\frac{5}{4}} & 0 & \cdot \\ \cdot & -\sqrt{\frac{5}{4}} & 0 & -1 & \cdot \\ \cdot & 0 & -1 & 0 & \cdot \\ \cdot & \cdot & \cdot & \cdot & 0 \end{array}} \end{array} \tag{37b}$$

3. BASIC STRUCTURE OF A MULTI-PARTICLE BASIS OF THE GELFAND REPRESENTATION

The mathematical structure of Gelfand's unitary representations can be explained from several viewpoints. Probably the best known approach involves a generalization of Schwinger's boson operator treatment of the quantum theory of angular momentum which Biedenharn and Louck [22] have developed. Here we shall sketch another approach to U_m started earlier by Weyl [23] and others, which uses permutation symmetry. The latter has turned out to be mathematically equivalent to the former, yet more convenient in some ways. Furthermore, we are motivated to treat carefully one of the apparantly perfect symmetries in nature; the identity of electrons.

It will now be shown how bases, such as the orbital states implied by Fig.8, are constructed using linear combinations of product state vectors (Eq.38a) or wave functions (Eq.38b).

$$|i_1\rangle^1 |i_2\rangle^2 \ \ldots \ |i_n\rangle^n = \left|\begin{matrix} 1 & 2 & \ldots n \\ i_1 & i_2 & \ldots i_n \end{matrix}\right\rangle \tag{38a}$$

$$\left\langle\begin{matrix} 1 & 2 & \ldots n \\ x_1 & x_2 & \ldots x_n \end{matrix}\middle|\begin{matrix} 1 & 2 & \ldots n \\ i_1 & i_2 & \ldots i_n \end{matrix}\right\rangle = \chi_{i_1}(x_1)\chi_{i_2}(x_2) \ \ldots \ \chi_{i_n}(x_n) \tag{38b}$$

$$\left\langle\begin{matrix} 1 & 2 & \ldots & n \\ i'_1 & i'_2 & \ldots & i'_n \end{matrix}\middle|\begin{matrix} 1 & 2 & \ldots & n \\ i_1 & i_2 & \ldots & i_n \end{matrix}\right\rangle = \delta_{i'_1 i_1} \ \delta_{i'_2 i_2} \ \ldots \ \delta_{i'_n i_n} \tag{38c}$$

(In Eq.38 the notation $\overset{\alpha}{i}$ or $\varphi_i(x_\alpha)$ indicates that particle α is in state i , and it is important that these states be orthonormal as per Eq.38c.) Now, because the particles are identical we must consider linear combinations of (n!) permutations of particles between states, in particular those combinations made with the well known permutation group (S_n) projectors $\bar{P}^\lambda_{\mu\nu}$ as in Eq.15 or Eq.39. There the factor $D^\lambda_{\mu\nu}(p)$ is an S_n representation component calculated using Fig.7, ℓ^λ is the dimension given by Fig.7-a, and N_ν is a normalization constant to be evaluated shortly.

$$|\mu,\nu\rangle = N_\nu \bar{P}^\lambda_{\mu\nu} \left| \begin{matrix} 1 & 2 & \dots & n \\ i_1 & i_2 & \dots & i_n \end{matrix} \right\rangle \tag{39}$$

$$= N_\nu (\ell^\lambda/n!) \sum_{\bar{p}} D^\lambda_{\mu\nu}(p)\, \bar{p} \left| \begin{matrix} 1 & 2 & \dots & n \\ i_1 & i_2 & \dots & i_n \end{matrix} \right\rangle$$

It turns out that a complete and orthonormal basis of any Gelfand representation can be made according to Eq.39, where $\bar{P}^\lambda_{\mu\nu}$ is applied only to those states $|i_1 i_2 \dots i_n\rangle$ of definite "order." By order we mean that single particle states have been numbered (Recall the numbering 1-3 in Sec. 2B) and these numbers are ordered: $i_1 \leq i_2 \leq \dots \leq i_n$. The permutations $\bar{p}$ in Eq.39 then give all other orderings. To help explain the properties of the permutation projectors (These properties were first discovered by Goddard[14].) we shall make an analogy with the quantum theory of rigid rotators.

The analogous rigid rotator state is made according to Eq.40 using rotation group (R_3) projectors where

$D^{\ell}_{mn}(\alpha\beta\gamma)$ is the rotation matrix for angular momentum ℓ, and the sum ("Σ") is an integral over Euler angles[24].

$$\left|{\ell \atop mn}\right\rangle = NP^{\ell}_{mn}|000\rangle = N''\sum_{\alpha\beta\gamma}{}'' D^{\ell *}_{mn}(\alpha\beta\gamma)\,\bar{O}(\alpha\beta\gamma)\,|000\rangle \tag{40}$$

In Eq.40 P^{ℓ}_{mn} has been applied to a state $|000\rangle$ of definite orientation (000). By (000) we mean that the rotator body axes 1, 2, and 3 are lined up with laboratory axes x, y, and z. The rotation operators $\bar{O}(\alpha\beta\gamma)$ in Eq.40 then give all other orientations. Now the quantum numbers m and n are eigenvalues of angular momentum operators J_z and J_3 for the lab z-axis and body 3-axis components respectively. In fact, two commuting groups of rotation operators $\bar{O}(\alpha\beta\gamma)$ and $\underline{O}(\alpha\beta\gamma)$ can be defined. $\bar{O}$ refers to the lab axis and rotates the body. $\underline{O}$ refers to the body axis and rotates the laboratory, and all $\underline{O}(\alpha\beta\gamma)$ commute with all $\bar{O}(\alpha'\beta'\gamma)$.

By analogy, two commuting groups of permutation operators $(\bar{p})$ and $(\underline{p})$ can be defined. A permutation $\bar{p}$ refers to the numbers on the electrons and permutes them from state to state. ($(\overline{124})$ means put electron (1) into whichever state that electron (2) was, electron (2) into whichever state that electron (4) was, and electron (4) into whichever state that electron (1) was.) A permutation $\underline{p}$ refers to the index numbers on the states and permutes these from electron to electron. ($(\underline{357})$ means put state-i_3 onto whichever electron state-i_5 was, state-i_5 onto whichever electron state-i_7 was, and state-i_7 onto whichever electron state-i_3 was.)

Therefore, (See Appendix A for detailed proof.) the "quantum numbers" or tableaus μ and ν in Eq.39 refer to the permutation symmetry of the state under permutation by $\bar{p}$ and $\underline{p}$ respectively.

Considering once again the corresponding quantum numbers m and n of the rigid rotator (Eq.40) we find that the ordinary z-component m varies over all $(2\ell+1)$ allowed values $-\ell \leq m \leq \ell$, while the body axis 3-component n may or may not be furthur restricted, depending on the nature of the body. If each rotation $\underline{O}(\alpha\beta\gamma)$ of the body gives a distinct position state then all values of n between $\pm\ell$ give valid states, too; m and n together label $(2\ell+1)(2\ell+1)$ states in total. However, if some subgroup of body rotations turn out to be just multiples of the identity, as would be rotations around the body axis of a diatomic molecule ($\underline{O}(\alpha 0 0)|\gamma\rangle = |\gamma\rangle$), then n may be restricted. In the latter example only n=0 survives, and the allowed rotator wave functions are just the common spherical harmonics $Y^{\ell}_{m} \sim D^{\ell *}_{m0}$.

There is an analogous freedom or restriction of the tableaus μ and ν for the multi-electron projection (Eq.39) where we find that μ varies over all ℓ^{2} lexical tableaus, while ν may or may not be furthur restricted depending on the nature of the unitary states $i_1 \leq i_2 \leq \ldots \leq i_n$. If each permutation $\underline{p}$ gives a distinct state (i.e. $i_\alpha \neq i_\beta$ for $\alpha \neq \beta$), then all lexical ν give valid states, too; μ and ν together label $\ell^{2} \cdot \ell^{2}$ states in total. (For example $3 \cdot 3 = 9$ orthogonal U_8 states are represented in Eq.41) However, if some subgroup of state permutations turn out to be just multiples of the identity as would be $(\underline{1})$ and $(\underline{12})$ for the state $|5567\rangle$ then ν may be restricted. (In the example only the 6 states represented in Eq.42 survive since $\nu = \begin{smallmatrix}134\\2\end{smallmatrix}$ can be shown to give zero, as explained in Appendix A.)

$$\left|\mu,\nu(i_\alpha) = \begin{matrix}235\\8\end{matrix}\right\rangle = N_1\, \bar{P}^{3100}_{\mu,\nu=\begin{smallmatrix}123\\4\end{smallmatrix}} \left|\begin{matrix}1234\\2358\end{matrix}\right\rangle \qquad (41)$$

$$\left|\mu,\begin{matrix}238\\5\end{matrix}\right\rangle = N_2\, \bar{P}^{3100}_{\mu\,\begin{smallmatrix}124\\3\end{smallmatrix}} \left|\begin{matrix}1234\\2358\end{matrix}\right\rangle$$

$$\left|\mu,\begin{matrix}258\\3\end{matrix}\right\rangle = N_3\, \bar{P}^{3100}_{\mu\,\begin{smallmatrix}134\\2\end{smallmatrix}} \left|\begin{matrix}1234\\2358\end{matrix}\right\rangle$$

$$\left|\mu,\begin{matrix}556\\7\end{matrix}\right\rangle = N_1'\, \bar{P}^{3100}_{\mu\,\begin{smallmatrix}123\\4\end{smallmatrix}} \left|\begin{matrix}1234\\5567\end{matrix}\right\rangle \qquad (42)$$

$$\left|\mu,\begin{matrix}557\\6\end{matrix}\right\rangle = N_2'\, \bar{P}^{3100}_{\mu\,\begin{smallmatrix}124\\3\end{smallmatrix}} \left|\begin{matrix}1234\\5567\end{matrix}\right\rangle$$

$$0 = \bar{P}^{3100}_{\mu\,\begin{smallmatrix}134\\2\end{smallmatrix}} \left|\begin{matrix}1234\\5567\end{matrix}\right\rangle$$

Eqs.41-42 show how unitary (U_m) tableaus $\nu(i_\alpha)$ are made from permutation (S_n) tableaus ν: the U_m state number i_α is put into the S_n tableau exactly where particle number α resides. Whenever two or more S_n tableaus ν, ν', ... give the same unitary tableau, it can be shown that the corresponding projections are identical except possibly for normalization.Eq.43 shows an example of this and further details are given in Appendix A.

$$\left| \mu, {}^{556}_{6} \right\rangle = N_1'' \, \bar{P}^{3100}_{\mu\, {}^{123}_{4}} \left| {}^{1234}_{5566} \right\rangle = N_2'' \, \bar{P}^{3100}_{\mu\, {}^{124}_{3}} \left| {}^{1234}_{5566} \right\rangle \tag{43}$$

The normalization constant for any of these states is given by Eq.44 as a sum over the identity permutations p^I that satisfy $p^I | i_1 \, i_2 \ldots \rangle = | i_1 \, i_2 \ldots \rangle$. This is derived in Appendix A.

$$1/N_\nu = \left[\frac{\ell^\lambda}{n!} \sum_{p^I} D^\lambda_{\nu\nu} (p^I) \right]^{\frac{1}{2}} \tag{44}$$

It can be shown that the preceeding formulation completely and unambiguously defines the bases for Gelfand representations which have been used in Secs. 1-2 . (See Appendix B.) However, the significance of the (μ) labels on these basis states needs to be clarified.

In all cases we find that for each allowed S_n tableau μ we get another complete basis for a U_m Gelfand representation defined by Young frame (λ). For different μ the representations of U_m come out exactly the same, but the bases are different, in fact, orthonormal.

Biedenharn and Louck produced Gelfand bases from polynomials of boson operators and thus avoided this repetition;[25]

in effect they keep only the "first tableau" μ_1, which has all the numbers in sequence 123... in every row. Certainly one basis per λ is all that is necessary as far as the mathematics of U_m is concerned. However, the explicit Pauli-antisymmetric states of orbit and spin (viz. the $U_3 \times U_2$ states implied by Eqs.29-30) involve a sum over all μ according to the formula in Eq.45. Each term in this sum includes an orbital factor of a particular μ with a spin factor of the transpose or conjugate tableau $\tilde{\mu}$ as seen in example Eqs.46. The first term has an orbit factor with the "first tableau" μ_1, and the spin factor with the "last tableau" $\tilde{\mu}_1$. (In $\tilde{\mu}_1$, all numbers are in sequence down the columns.)
Each following term has in the orbit factor, an allowed μ which is a permutation of numbers in μ_1, and this term is positive (negative) if this permutation is even (odd).

(45)

$$\left|\nu(i_\alpha)\ \ \nu'(s_\alpha)\right\rangle = \sum_{\mu_j=\mu_1}^{j=\ell^\lambda} (\pm 1)/\sqrt{\ell^\lambda}\ \left(N_\nu\ P^{\lambda}_{\mu_j \nu} \left|\begin{matrix}1 & 2 & \cdots \\ i_1 & i_2 & \cdots\end{matrix}\right\rangle\right)\left(N_{\nu'}\ P^{\tilde{\lambda}}_{\tilde{\mu}_j \nu'} \left|\begin{matrix}1 & 2 & \cdots \\ s_1 & s_2 & \cdots\end{matrix}\right\rangle\right)$$

$$= \sum_{\mu_j=\mu_1}^{j=\ell^\lambda} (\pm 1)/\sqrt{\ell^\lambda}\ \left|(\mu_j),\ \nu(i_\alpha)\right\rangle \left|(\tilde{\mu}_j)\ \ \nu'(s_\alpha)\right\rangle$$

(46)

$$\left|\begin{matrix}56 & \uparrow\uparrow\downarrow \\ 6 & \downarrow \\ 7 & \end{matrix}\right\rangle = 1/\sqrt{3}\left|\mu = \begin{matrix}12\\3\\4\end{matrix},\ \nu(i_\alpha) = \begin{matrix}56\\6\\7\end{matrix}\right\rangle\left|\tilde{\mu} = \begin{matrix}134\\2\end{matrix},\ \nu'(s_\alpha) = \begin{matrix}\uparrow\uparrow\downarrow\\ \downarrow\end{matrix}\right\rangle$$

$$- 1/\sqrt{3}\left|\begin{matrix}13\\2\\4\end{matrix},\ \begin{matrix}56\\6\\7\end{matrix}\right\rangle\left|\begin{matrix}124\\3\end{matrix},\ \begin{matrix}\uparrow\uparrow\downarrow\\ \downarrow\end{matrix}\right\rangle$$

$$+ 1/\sqrt{3}\left|\begin{matrix}14\\2\\3\end{matrix},\ \begin{matrix}56\\6\\7\end{matrix}\right\rangle\left|\begin{matrix}123\\4\end{matrix},\ \begin{matrix}\uparrow\uparrow\downarrow\\ \downarrow\end{matrix}\right\rangle$$

$$\left|\begin{matrix}56 & \uparrow\uparrow \\ 67 & \downarrow\downarrow\end{matrix}\right\rangle = 1/\sqrt{2}\left|\begin{matrix}12 & & 56 \\ 34 & , & 67\end{matrix}\right\rangle\left|\begin{matrix}13 & & \uparrow\uparrow \\ 24 & , & \downarrow\downarrow\end{matrix}\right\rangle$$

$$- 1/\sqrt{2}\left|\begin{matrix}13 & & 56 \\ 24 & , & 67\end{matrix}\right\rangle\left|\begin{matrix}12 & & \uparrow\uparrow \\ 34 & , & \downarrow\downarrow\end{matrix}\right\rangle$$

It is undeed fortunate that most of this complexity does not have to be brought out again when this mathematics is used for physical problems. In this respect the notation and accompanying calculus is most efficient.

However, the proofs of tableau formulas and the correspondences between Gelfand and tableau approaches[26] have been quite difficult to make, possibly because the former is algebraic while the latter is not. The nature of this dilemma is seen in the proofs and discussion here, in Part II, and in works by others.[27] We hope that soon there will exist less complicated ways to derive such simple results.

PART II: STRUCTURE AND APPLICATIONS

Modern applications or studies of atomic or molecular physics usually involve the quantum mechanics of many states and particles. The unitary calculus described in Part I (Henceforward we refer to this as (I)) is designed to treat such complexity very efficiently. Nevertheless, we expect that one will generally prefer to use computing machines to perform calculations, and so it is important that applications of the calculus be explained in a straightforward way that can be coded.

The application of the unitary formulation of angular factor analysis of complex atomic spectra is described below in Sec. 1. Here we find one set of formulas and rules that are the same for all configurations; pure or mixed, and no matter how many sub-shells or electrons are involved. We explain a way to take advantage of rotational symmetry or angular momentum conservation while dealing efficiently with independent terms of equal total orbital momentum and spin multiplicity. This is done without involving the baroque "lores" of the various Racah bases which are necessarily quite difficult to teach to a computing machine. (However it is interesting to relate tableau bases to Racah states as seen in Sec. 1.) Instead we always use the same unitary bases of (I) whose structure is given mainly by two formulas (Fig.9 and Fig.10) and whose correspondence with a given physical model is never ambiguous. We discuss the physical models involving L-S, intermediate, and jj-coupling approximations and their interrelations for nearly free atoms.

The ways of taking advantage of crystal or molecular point symmetry is seen in Sec. 2, where some applications to the study of molecular electronic structure are shown. Here again, the same basic formulas and rules apply universally, however, the possibilities for different physical models and approximation schemes is much richer.

For either of these applications it seems to be better to have bases which are defined with respect to spatial symmetries only after they are defined with respect to: (1) the permutation symmetry S_n for n electrons, and (2) the complete operator group U_m defined for m states. Indeed, this S_n symmetry is always quite perfect in nature while the various spatial symmetries come and go, and strictly speaking, can never be perfect in the real world. And, the operator group U_m is simply a mathematical prerequisite for doing quantum mechanics with m states when their total probability is to be conserved.

The last Section 3 is devoted to further discussion of the mathematical structure of S_n and U_m bases, i.e. the pattern calculus in general. There we discuss the parentage and shell structures which the patterns make quite obvious, and give a more general procedure for dealing with particles of higher spin or more internal structure than the non-relativistic electron. This theory has led to the tableau formulas of (I) and is certainly capable of producing more general analyses of states involving nucleons, hadrons, or relativistic models.

1. UNITARY ANALYSIS FOR NEARLY-FREE ATOMS

The unitary analysis of atomic Russell-Saunders (LS) coupling states was introduced in (I) and will be treated in more detail now. The atomic p sub-shell which was used as an example in (I) does not exhibit all the structural problems that arise in general, namely the appearance of independent terms of identical ^{2S+1}L and the treatment of multiple shells or "mixed configurations." We now treat the simplest examples that do exhibit these problems, and give routines for obtaining the energy and transition matrices using the least number of mathematical operations.

The routines for treating multiple shells are only slightly different from those that treat pure configurations, so most of the procedural details are are contained in the discussion of the latter which is given next below.

A. PURE ORBITAL CONFIGURATIONS $(\ell)^m$

The pure configuration or single (ℓ) sub-shell problem involves $(2\ell+1)$ single electron orbital states which we shall label $|1\rangle = |\ell\rangle$, $|2\rangle = |\ell-1\rangle$, ... $|2\ell+1\rangle = |-\ell\rangle$. The combinations $E_{ij} = |i\rangle\langle j|$ span the $U_{2\ell+1}$ operator algebra. Examples of relations between the unit multipole operators V^k_q and the elementary operators E_{ij} were given for the p sub-shell by Eq.16-17 in (I). There the coefficients relating the two were just the conventional matrix elements of V^k_q as given by Eq.1 below. (See also Eqs. (2), (10), and (11) in (I).)

$$V^k_q = \sum_{m\, m'} \left|{\ell \atop m}\right\rangle\left\langle{\ell \atop m}\right| V^k_q \left|{\ell \atop m'}\right\rangle\left\langle{\ell \atop m'}\right| = \sum_{m\, m'} \left\langle{\ell \atop m}\right| V^k_q \left|{\ell \atop m'}\right\rangle E_{(m)(m')} \tag{1a}$$

$$\left\langle{\ell \atop m}\right| V^k_q \left|{\ell \atop m'}\right\rangle = (-1)^{\ell - m}\sqrt{2k+1}\begin{pmatrix} \ell & k & \ell \\ -m & q & m' \end{pmatrix} = C^{k\ \ell\ \ell}_{q\ m'\ m}(-1)^k\left(\frac{2k+1}{2\ell+1}\right)^{\frac{1}{2}} \tag{1b}$$

Table III gives these V^k_q for ℓ = 1-3 and 4 respectively, i.e. for p, d, f, and g sub-shells. Note that matrices for V^k_q with k and ℓ fixed but with various $q(-k \leqslant q \leqslant k)$ are superimposed since each V^k_q matrix is zero except along one particular row parallel to the diagonal. The lowest common denominators of each V^k_q are printed on the right of each matrix in line with its particular super diagonal row. (Note also that $\tilde{V}^k_q = (-1)^q V^k_{-q}$.) For example, V^6_4 is given below from Table III-f.

$$V^6_4 = (\sqrt{5}\, E_{15} - \sqrt{12}\, E_{26} + \sqrt{5}\, E_{37})/\sqrt{22} \tag{2}$$

For fixed (ℓ, q) there are $(2\ell - q + 1)$ operators V^k_q $(k \geqslant q)$ that have this same number of components in the same positions of the matrix. The definition in Eq.1 is made so all these diagonal rows form orthonormal sets of vectors. Therefore the inverses of V - E relations can be found by inspection. (Viz. Eq.3 from Table III-f.)

$$E_{15} = \sqrt{\frac{5}{22}}\, V^6_4 + \frac{1}{\sqrt{2}}\, V^5_4 + \sqrt{\frac{3}{11}}\, V^4_4 \tag{3}$$

For the analysis of free atoms one needs the expectation values of matrices of operators proportional to V^k_q and $\left(V^k \cdot V^k\right)$, between given terms ^{2S+1}L allowed by the configuration under study. We now show how to accomplish this in a routine manner.

First all the spectral-term states of interest must be extracted. This is complicated slightly when more than one term with the same L and S show up, i.e. when repeated L arise with a given spin-multiplicity. The simplest example of a single sub-shell multiplet with two equal L terms is found in the doublets of the $(d)^3$ configuration. The program we now apply to this will, in principle, solve any atomic configuration.

The program is initialized by writing the highest tableau of the multiplet in question, where the shape of the frame follows from the multiplicity (See Fig.1), and the numbers go into it as follows: $\begin{smallmatrix}11\\22\\33\\ \cdot\cdot\\ \cdot\cdot\\ \cdot\end{smallmatrix}$. For $(d)^3$ doublets we start with $\begin{smallmatrix}11\\2\end{smallmatrix}$.

Then all the U_m $(m=2\ell+1)$ "lowering operators" E_{21}, $E_{32}, E_{43}, \ldots, E_{m\,m-1}$ are applied once to the initial tableau. The new tableaus that result are transcribed along with the matrix elements (Recall Fig.9) for the change. Only tableaus that have the numbers equal or increasing toward the right in their rows, and increasing but <u>not</u> equal down their columns, are transcribed. For $(d)^3$ doublets we obtain the tableaus $\begin{smallmatrix}12\\2\end{smallmatrix}$ and $\begin{smallmatrix}11\\3\end{smallmatrix}$ with matrix elements $\left\langle\begin{smallmatrix}12\\2\end{smallmatrix}\right| E_{21} \left|\begin{smallmatrix}11\\2\end{smallmatrix}\right\rangle$ $=1$ and $\left\langle\begin{smallmatrix}11\\3\end{smallmatrix}\right| E_{32} \left|\begin{smallmatrix}11\\2\end{smallmatrix}\right\rangle$ $=1$ respectively.

This procedure is repeated upon each of the new tableaus to make a third set, and then a fourth, and so forth. Each stage gives a complete set states with a fixed total-M, thereby indicating directly what ^{2S+1}L terms are allowed.

Now as the lowering operators are applied at each stage, one should transcribe matrix elements for all allowed lowerings because, among other things, they serve to sort out the ^{2S+1}L states. For example, component $\left\langle \begin{smallmatrix}12\\3\end{smallmatrix} \middle| E_{32} \middle| \begin{smallmatrix}12\\2\end{smallmatrix} \right\rangle = \sqrt{\tfrac{1}{2}}$ should be recorded even though the tableau $\begin{smallmatrix}12\\3\end{smallmatrix}$ would already be made by applying E_{21} to $\begin{smallmatrix}11\\3\end{smallmatrix}$. ($\left\langle \begin{smallmatrix}12\\3\end{smallmatrix} \middle| E_{21} \middle| \begin{smallmatrix}11\\3\end{smallmatrix} \right\rangle = \sqrt{2}$)

In Eq.5 the matrix elements of the raising operators $E_{m-1\ m}$ (These are simply reflections through the diagonal of $E_{m\ m-1}$ components.) are transcribed and labeled above the diagonal. The components of the angular momentum lowering operator L_- (Eq.4) are written below the diagonal in Eq.5.

$$L_- \equiv \sqrt{20}\, V^1_{-1} = 2E_{21} + \sqrt{6}\, E_{32} + \sqrt{6}\, E_{43} + 2E_{54} \tag{4}$$

For any atomic configuration the highest tableau always corresponds to the highest L term in the multiplet with z-component of orbit M=L. For the example $\left| \begin{smallmatrix}11\\2\end{smallmatrix} \right\rangle = \left| ^2H\ L=5,\ M=5 \right\rangle$. Now an application of L_-,according to Eq.6 and an orthogonalization which is described below, gives all the other terms.

$$\frac{L_-}{\sqrt{(L+M)(L-M+1)}} \left| L,M \right\rangle = \left| L,M-1 \right\rangle \tag{6}$$

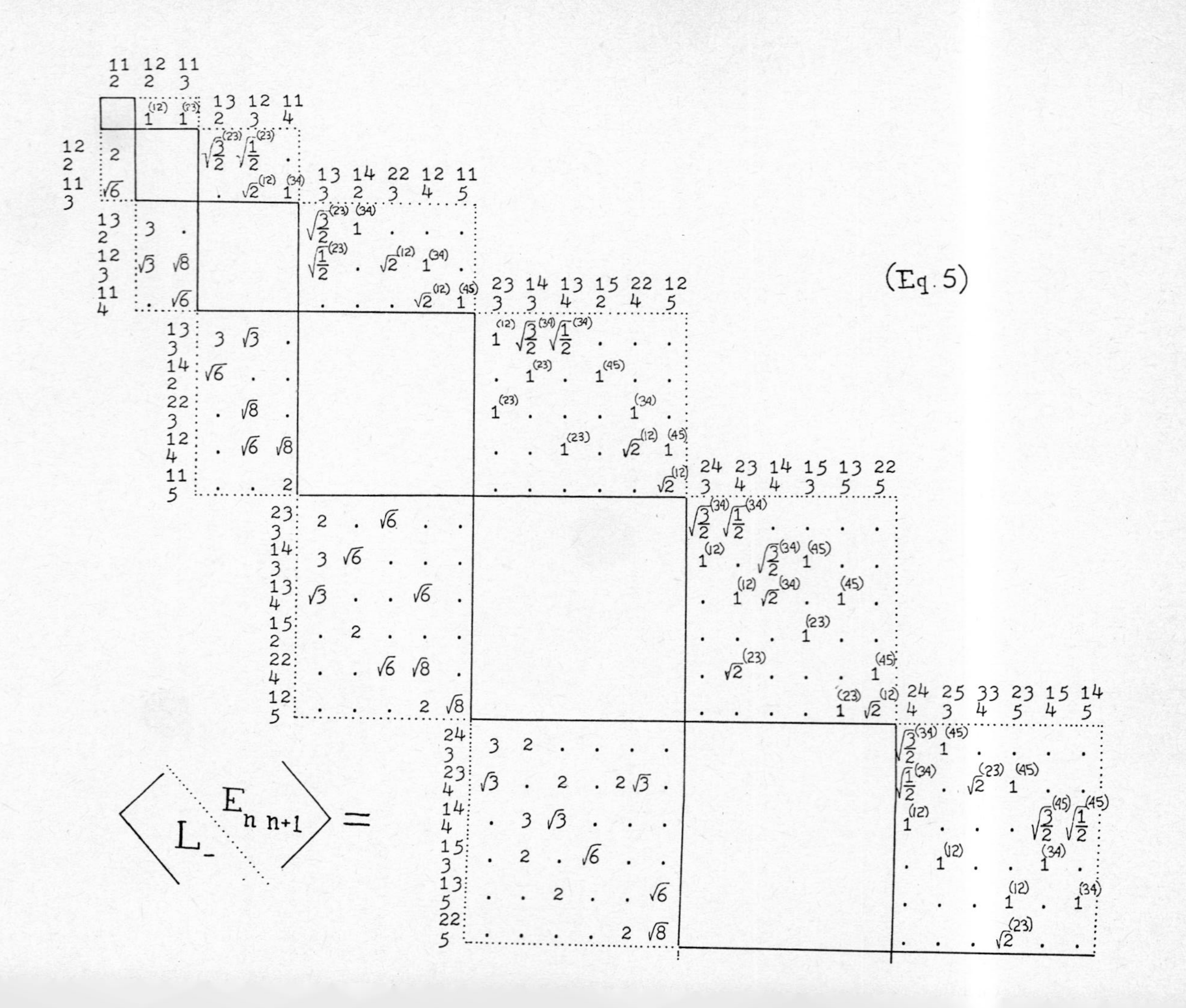
(Eq. 5)

For the example we obtain the $|{}^2G\ 4,4\rangle$ state (Eq.7b) from orthogonalization with the $|{}^2H\ 5,4\rangle$ state (Eq.7a) and similarly for the next set of M=3 states in Eq.8.

$$|{}^2H\ 5,4\rangle = \frac{L_-}{\sqrt{10}}\,|{}^2H\ 5,5\rangle = \left(2 \genfrac{}{}{0pt}{}{12}{2} + \sqrt{6} \genfrac{}{}{0pt}{}{11}{3}\right) / \sqrt{10} \tag{7a}$$

$$|{}^2G\ 4,4\rangle = \left(\sqrt{6} \genfrac{}{}{0pt}{}{12}{2} - 2 \genfrac{}{}{0pt}{}{11}{3}\right) / \sqrt{10} \tag{7b}$$

$$|{}^2H\ 5,3\rangle = \frac{L_-}{\sqrt{18}}\,|{}^2H\ 5,4\rangle = \left(\genfrac{}{}{0pt}{}{13}{2} + \sqrt{3} \genfrac{}{}{0pt}{}{12}{3} + \genfrac{}{}{0pt}{}{11}{4}\right) / \sqrt{5} \tag{8a}$$

$$|{}^2G\ 4,3\rangle = \frac{L_-}{\sqrt{8}}\,|{}^2G\ 4,4\rangle = \left(3\sqrt{3} \genfrac{}{}{0pt}{}{13}{2} - \genfrac{}{}{0pt}{}{12}{3} - 2\sqrt{3} \genfrac{}{}{0pt}{}{11}{4}\right) / \sqrt{40} \tag{8b}$$

$$|{}^2F\ 3,3\rangle = \left(\genfrac{}{}{0pt}{}{13}{2} - \sqrt{3} \genfrac{}{}{0pt}{}{12}{3} + 2 \genfrac{}{}{0pt}{}{11}{4}\right) / \sqrt{8} \tag{8c}$$

However, we will need a rigorous and efficient routine which can produce repeated orbit states as well as the simple unrepeated states in Eq.7b or Eq.8c. For example there will be two terms ${}^2D$ and ${}^2D'$ needed to complete the next set of five M=2 states of which only the three in Eq.9 are found using L_-.

$$\frac{L_-}{\sqrt{24}}\,|{}^2H\ 5,3\rangle = |{}^2H\ 5,2\rangle = \left(6 \genfrac{}{}{0pt}{}{13}{3} + \sqrt{6} \genfrac{}{}{0pt}{}{14}{2} + 2\sqrt{6} \genfrac{}{}{0pt}{}{22}{3} + 5\sqrt{2} \genfrac{}{}{0pt}{}{12}{4} + 2 \genfrac{}{}{0pt}{}{11}{5}\right) / 2\sqrt{30} \tag{9a}$$

$$|{}^2G\ 4,2\rangle = \left(4\sqrt{6} \qquad 9 \qquad -2 \qquad -5\sqrt{3} \qquad -2\sqrt{6}\right) / 2\sqrt{70} \tag{9b}$$

$$|{}^2F\ 3,2\rangle = \left(0 \qquad \sqrt{3} \qquad -2\sqrt{3} \qquad 1 \qquad 2\sqrt{2}\right) / 2\sqrt{6} \tag{9c}$$

An orthonormal set of bases can be made using the submatrix $\langle S \rangle$ of projection operator S in Eq.10. The sum there is over the equal M vectors previously derived such as those in Eq.9.

$$S = 1 - \sum_{\substack{L>M \\ M \text{ fixed}}} |L , M\rangle\langle L , M| \tag{10}$$

Certain properties of the S operator minimize the computational labor. First, $\langle S \rangle$ is a symmetric projector so its rows or its columns are the components of the desired eigenvectors of L^2. Secondly, S is idempotent ($S \cdot S = S$) so the scalar product $(i|j)$ of the i-th row vector $(i|$ with the j-th column vector $|j)$ must equal the i-j entry S_{ij} of the matrix $\langle S \rangle$. We denote vectors that may not be orthonormal by modified Dirac symbols $|\ )$ or $(\ |$.

Therefore all the quantities needed for Gram-Schmidt[28] orthogonalization are components of the matrix $\langle S \rangle$. Let us define $|I) = |i)$ using the first column-i having a non-zero norm $(i|i)$. (The i-th column is all zero if and only if its diagonal element $(i|i)$ is zero.) The normalized $|I\rangle$ (Eq.11a) will be taken as the first of the desired angular momentum state vectors, as is the example in Eq.11b.

$$|I\rangle = |I) /\sqrt{(I|I)} = |i) /\sqrt{(i|i)} \tag{11a}$$

(11b)

$$|\,I\rangle = |\,d^3\;{}^2D\;2,2\rangle = \left(\frac{5}{14}\,{}^{13}_{3} - \frac{15}{14\sqrt{6}}\,{}^{14}_{2} - \frac{3}{7\sqrt{6}}\,{}^{22}_{3} - \frac{1}{14\sqrt{2}}\,{}^{12}_{4} + \frac{1}{14}\,{}^{11}_{5}\right)\Big/\sqrt{\frac{5}{14}}$$

$$= (5\sqrt{2} \quad -5\sqrt{3} \quad -2\sqrt{3} \quad -1 \quad \sqrt{2}\;)/\sqrt{140}$$

Now it must be found if the next column $|j) = |i+1)$ gives a vector independent of $|I)$. A simple test routine exists which will check this by calculating just one component. If the result is null, the routine must try the following columns $|j) = |i+2)$, $|i+3)$, ... until a non-null result is found. A good thing about the routine is that <u>the number of calculations needed to obtain an orthonormal vector is constant</u> no matter how far the test goes before getting a non-null result.

To see this, note that the vector $|II)$ in Eq.12a will be orthogonal to $|I)$, and suitable as the second angular momentum state if its norm (Eq.12b) is non-zero. However, this norm, if non-zero, is seen to be the j-th component of $|II)$, (Eq.12c) i.e. the nullity test result becomes a vector component and the rest of the vector is constructed according to Eq.12a, using the j=m-th column. For each failure of the test for j we find that j-th component of $|II)$ is identically zero, as are all components preceeding the first non-null result.

$$|II) = |j) - \frac{|I)(I|j)}{(I|I)} \tag{12a}$$

$$(II|II) = \left((j| - \frac{(j|I)(I|}{(I|I)}\;,\; |j) - \frac{|I)(I|j)}{(I|I)}\right) = (j|j) - \frac{(I|j)^2}{(I|I)} \tag{12b}$$

$$(\mathrm{II}|\mathrm{II}) = (j|\mathrm{II}) \tag{12c}$$

For the ^{2}D state problem one finds that the $j=2$ column test is null while the $j=3$ column gives the orthonormal vector in Eq.13 to complete the sets in Eq.9a-c and Eq.11b.

$$|\mathrm{II}\rangle = |\mathrm{II})/\sqrt{(\mathrm{II}|\mathrm{II})} = |d^3\ {}^2D'\ 2,2\rangle \tag{13}$$

$$= \left(\frac{1}{5}\begin{matrix}22\\3\end{matrix} - \frac{2}{5\sqrt{3}}\begin{matrix}12\\4\end{matrix} + \frac{4}{5\sqrt{6}}\begin{matrix}11\\5\end{matrix} \right) / \sqrt{\frac{1}{5}}$$

$$= (0 \quad 0 \quad \sqrt{6} \quad -2\sqrt{2} \quad 4 \quad)/\sqrt{30}$$

Basically the same procedure is applied again and again if the number of tableaus dictate that such is necessary, (Our example terminates with $|\mathrm{II})$ since there were only two more tableaus in the $M=2$ set than in the $M=3$ set.) The next orthogonal vector would be $|\mathrm{III})$ defined in Eq.14 using the k-th column, and again the nullity test or norm of $|\mathrm{III})$ *is* also its k-th component.

$$|\mathrm{III}) = |k) - \frac{|\mathrm{I})(\mathrm{I}|k)}{(\mathrm{I}|\mathrm{I})} - \frac{|\mathrm{II})(\mathrm{II}|k)}{(\mathrm{II}|\mathrm{II})} \tag{14a}$$

$$(\mathrm{III}|\mathrm{III}) = (k|k) - \frac{(k|\mathrm{I})^2}{(\mathrm{I}|\mathrm{I})} - \frac{(k|\mathrm{II})^2}{(\mathrm{II}|\mathrm{II})} = (k|\mathrm{III}) \tag{14b}$$

Always the result of a nullity test (14b) becomes the k-th component of the new vector. If that component is zero, Eq.14b is used again with $k \rightarrow k+1$ to get the next (k+1-th) component. If it is non-zero, the remaining components come from Eq.14a or Eq.15 below.

$$(m|III) = (m|k) - \frac{(m|I)(I|k)}{(I|I)} - \frac{(m|II)(II|k)}{(II|II)} \tag{15}$$

This routine uses a minimum and fixed number of components $(m|k)$ of the submatrix $\langle S \rangle$ in Eq.10 to arrive at a set of angular momentum eigenstates. The "true" set of eigenstates will be mixtures within each set found by diagonalizing the electrostatic repulsion operator (E.R.) in Eq.16, and the result depends upon the values of the radial parameters f_k or equivalently, the Slater parameters F^k.

$$(E.R.) = \sum_{\substack{k=0 \\ (\text{even}) \\ k}}^{2\ell} f_k \, (V^k \cdot V^k) \tag{16}$$

A more detailed discussion of these parameters for the electrostatic interaction is found at the end of this section and in Sec. 1B.

The final orthonormalized form of the repeated ^{2S+1}L state vectors $|I\rangle$, $|II\rangle$, $|III\rangle$, ... facilitates the calculation of the interaction matrices of the $V^k \cdot V^k$ operators in those bases. The positions of the zeros indicated in Eq.17 make it easy to find the desired components with a minimum amount of matrix algebra.

$$
\begin{aligned}
|I\rangle &= 0 \ldots \langle i|I\rangle \ldots\ldots \langle j|I\rangle \ldots \langle k|I\rangle \ldots \\
|II\rangle &= 0 \ldots \quad 0 \quad \ldots\ldots \langle j|II\rangle \ldots \langle k|II\rangle \ldots \\
|III\rangle &= 0 \ldots \quad 0 \quad \ldots\ldots \quad 0 \quad \ldots \langle k|III\rangle \ldots
\end{aligned}
\tag{17}
$$

Suppose, as before, that the tableaus or components i, j, k, ... corresponded to the first non-zero components of $|I\rangle$, $|II\rangle$, $|III\rangle$, ... respectively. Then the calculation of the desired matrix components $\langle I|V^k \cdot V^k|I\rangle$, $\langle I|V^k\, V^k|II\rangle$ ... will require only the i, j, k, ... rows (Eq.18) of the submatrix $\langle i|V^k \cdot V^k|j\rangle$ in the basis of the equal-M tableaus $|1\rangle$, $|2\rangle$, ... $|n\rangle$.

$$
\langle V^k \cdot V^k\rangle = \begin{matrix}
\langle i|V^k \cdot V^k|1\rangle & \langle i|V^k \cdot V^k|2\rangle & \ldots & \langle i|V^k \cdot V^k|n\rangle \\
\langle j|V^k \cdot V^k|1\rangle & \langle j|V^k \cdot V^k|2\rangle & \ldots & \langle j|V^k \cdot V^k|n\rangle \\
\langle k|V^k \cdot V^k|1\rangle & \langle k|V^k \cdot V^k|2\rangle & \ldots & \langle k|V^k \cdot V^k|n\rangle
\end{matrix}
\tag{18}
$$

The product of the rectangular matrix in Eq.18 with the rectangular matrix made from Eq.17 gives a square matrix of quantities (Eq.19a) that are easily converted into the interaction matrix through Eq.19b.

$$\langle v^k \cdot v^k \rangle \; \begin{matrix} |I\rangle & |II\rangle & |III\rangle & \cdots \end{matrix} = \begin{matrix} h_{11} & h_{12} & h_{13} & \cdots \\ h_{21} & h_{22} & h_{23} & \cdots \\ h_{31} & h_{32} & h_{33} & \cdots \\ \vdots & \vdots & \vdots & \end{matrix} \qquad (19\text{-a})$$

$$\langle I|V|I\rangle = \frac{h_{11}}{\langle i|I\rangle} \qquad \langle I|V|II\rangle = \frac{h_{12}}{\langle i|I\rangle} \qquad \langle I|V|III\rangle = \frac{h_{13}}{\langle i|I\rangle}$$

$$(19\text{-b})$$

$$\langle II|V|I\rangle = \langle I|V|II\rangle \qquad \langle II|V|II\rangle = \frac{h_{22}}{\langle j|II\rangle} - \frac{\langle I|V|II\rangle\langle j|I\rangle}{\langle j|II\rangle} \qquad \langle II|V|III\rangle = \frac{h_{23}}{\langle j|II\rangle} - \frac{\langle I|V|III\rangle\langle j|I\rangle}{\langle j|II\rangle}$$

$$\langle III|V|I\rangle = \langle I|V|III\rangle \qquad \langle III|V|II\rangle = \langle II|V|III\rangle \qquad \langle III|V|III\rangle = \frac{h_{33}}{\langle k|III\rangle} - \frac{\langle I|V|III\rangle\langle k|I\rangle - \langle II|V|III\rangle\langle k|II\rangle}{\langle k|III\rangle}$$

where: $V \equiv v^k \cdot v^k$

The components $\langle i | V^k \cdot V^k | j \rangle$ in Eq.18 of the operator in Eq.20 for the tableau basis are calculated by first expressing the V^k_q in terms of the elementary operators as was done by Eq.1 or the Tables II-III.

$$V^k \cdot V^k = \sum_{q=-k}^{k} (-1)^q \, V^k_{-q} \, V^k_q = \sum_{q=-k}^{k} \tilde{V}^k_q \, V^k_q \tag{20}$$

$$= (V^k_0)^2 + \sum_{q=1}^{k} (\tilde{V}^k_q \, V^k_q + V^k_q \, \tilde{V}^k_q)$$

These relations are expressed again in Eq.21 using a more streamlined notation (mn) for the coefficients. In machine computation these coefficients could be calculated when needed or brought from memory, whichever is more convenient. The latter is quite feasible since the number of them, after using obvious symmetry relations, (See Tables II-III) is not great. The f sub-shell requires only 69 different numbers, for example.

$$V^k_o = \sum_m (m \; m) \, E_{m \, m} \tag{21}$$

$$V^k_q = \sum_m (m \;\; m+q) \, E_{m \;\, m+q}$$

$$\tilde{V}^k_q = \sum_n (n \;\; n+q) \, E_{n+q \;\, n}$$

The preceeding two equations give the elementary definition of the multipole interaction operator in Eq.22.

$$V^k \cdot V^k = (\Sigma (m\ m)\ E_{m\ m})^2 \tag{22}$$

$$+ \sum_{q=1}^{k} (\sum_m \sum_n (m\ m{+}q)(n\ n{+}q)(E_{m\ m+q}\ E_{n+q\ n} + E_{n+q\ n}\ E_{m\ m+q})$$

As an example, we shall evaluate these operators for the d^3, ^2D and $^2D'$ states. To do this we need only two rows of the 5x5 matrix $V^k \cdot V^k$ in the tableau basis, namely those headed by the first $\binom{13}{3}$ and the third $\binom{22}{3}$ tableau states. These are written out below (Eqs.23) with the first two showing explicitly which operators contributed.

$$\left\langle {13 \atop 3} \middle| V^k \cdot V^k \middle| {13 \atop 3} \right\rangle = \left\langle {13 \atop 3} \middle| ((11)\ E_{11} + (33)\ E_{33})^2 \middle| {13 \atop 3} \right\rangle \tag{23}$$

$$+ \left\langle {13 \atop 3} \middle| (12)^2 (E_{12}\ E_{21}) + (23)^2 (E_{32}\ E_{23}) + (34)^2 (E_{34}\ E_{43}) \middle| {13 \atop 3} \right\rangle$$

$$+ \left\langle {13 \atop 3} \middle| (13)^2 (\qquad E_{31}\ E_{13}) + (35)^2 (E_{35}\ E_{53}) \middle| {13 \atop 3} \right\rangle$$

$$+ \left\langle {13 \atop 3} \middle| (14)^2\ E_{14}\ E_{41} \middle| {13 \atop 3} \right\rangle$$

$$+ \left\langle {13 \atop 3} \middle| (15)^2\ E_{15}\ E_{51} \middle| {13 \atop 3} \right\rangle$$

$$\left\langle {13 \atop 3} \middle| V^k \cdot V^k \middle| {13 \atop 3} \right\rangle = ((11) + 2(33))^2 + (12)^2 + 2(23)^2 + 2(34)^2$$

$$+ (13)^2 + 2(35)^2 + (14)^2 + (15)^2$$

$$\left\langle {13 \atop 3} \middle| V^k \cdot V^k \middle| {14 \atop 2} \right\rangle = \left\langle {13 \atop 3} \middle| (23)(34)(E_{34}\ E_{32} + E_{32}\ E_{34}) \middle| {14 \atop 2} \right\rangle$$

$$= (23)(34)\sqrt{6}$$

$$\left\langle \begin{smallmatrix}13\\3\end{smallmatrix} \middle| V^k\cdot V^k \middle| \begin{smallmatrix}22\\3\end{smallmatrix} \right\rangle = 2(12)(23) \tag{23}$$

$$\left\langle \begin{smallmatrix}13\\3\end{smallmatrix} \middle| V^k\cdot V^k \middle| \begin{smallmatrix}12\\4\end{smallmatrix} \right\rangle = (23)(34)\sqrt{2}$$

$$\left\langle \begin{smallmatrix}13\\3\end{smallmatrix} \middle| V^k\cdot V^k \middle| \begin{smallmatrix}11\\5\end{smallmatrix} \right\rangle = 2(13)(35)$$

$$\left\langle \begin{smallmatrix}22\\3\end{smallmatrix} \middle| V^k\cdot V^k \middle| \begin{smallmatrix}14\\2\end{smallmatrix} \right\rangle = -(13)(24)\sqrt{6}$$

$$\left\langle \begin{smallmatrix}22\\3\end{smallmatrix} \middle| V^k\cdot V^k \middle| \begin{smallmatrix}22\\3\end{smallmatrix} \right\rangle = (2(22) + (33))^2 + 2(12)^2 + (23)^2 + (34)^2 + (13)^2 + 2(24)^2 + (35)^2 + 2(25)^2$$

$$\left\langle \begin{smallmatrix}22\\3\end{smallmatrix} \middle| V^k\cdot V^k \middle| \begin{smallmatrix}12\\4\end{smallmatrix} \right\rangle = 2(12)(34)\sqrt{2} - (13)(24)\sqrt{2}$$

$$\left\langle \begin{smallmatrix}22\\3\end{smallmatrix} \middle| V^k\, V^k \middle| \begin{smallmatrix}11\\5\end{smallmatrix} \right\rangle = 0$$

It is instructive to see how these components and others like them can be read by inspection of selected parts of the representation of the E_{ij}'s shown in Eq.24. This turns out to be a quite acceptable way to do calculations by hand. Producing Eq.24 from Eq.5 is quite simple using the basic relations of the generators given by Eq.18 in (I), and provides a continuous check of the results along the way. Now, for example, the second to last matrix element in Eq.23 is obtained from Eq.24 by adding the overlap between the $\begin{smallmatrix}22\\3\end{smallmatrix}$ and $\begin{smallmatrix}12\\4\end{smallmatrix}$ columns to the overlap between the same two rows.

A numerical evaluation of a particular $V^k\cdot V^k$, say $V^3\cdot V^3$ for example, proceeds after substituting the (m n)

$$\langle E_{mn} \rangle =$$

M		11 2	12 2	11 3	13 2	12 3	11 4	13 3	14 2	22 3	12 4	11 5	23 3	14 3	13 4	15 2	22 4	12 5	24 3	23 4	14 4	15 3	13 5	22 5	24 4	25 3	33 4	23 5	15 4	14 5	34 4	25 4	24 5	33 5	15 5
M=5	11 2		$1^{(12)}$	$1^{(23)}$	$\sqrt{\frac{3}{2}}^{(13)}$	$\frac{\sqrt{1}}{\sqrt{2}}^{(13)}$	$1^{(24)}$	.	$\sqrt{\frac{3}{2}}^{(14)}$	.	$\frac{\sqrt{1}}{\sqrt{2}}^{(14)}$	$1^{(25)}$	.	.	.	$\sqrt{\frac{3}{2}}^{(15)}$	.	$\frac{\sqrt{1}}{\sqrt{2}}^{(15)}$																	
M=5,4	12 2				$\sqrt{\frac{3}{2}}^{(23)}$	$\sqrt{\frac{1}{2}}^{(23)}$	.	.	$\sqrt{\frac{3}{2}}^{(24)}$	$-1^{(13)}$	$\sqrt{\frac{1}{2}}^{(24)}$	.	.	.	.	$\sqrt{\frac{3}{2}}^{(25)}$	$-1^{(14)}$	$\sqrt{\frac{1}{2}}^{(25)}$	.	.	.	.	.	$-1^{(15)}$											
	11 3				.	$\sqrt{2}^{(12)}$	$1^{(34)}$	$1^{(13)}$	.	.	.	$1^{(35)}$	.	$\sqrt{\frac{3}{2}}^{(14)}$	$\frac{\sqrt{1}}{\sqrt{2}}^{(14)}$	.	.	.	.	.	.	$\sqrt{\frac{3}{2}}^{(15)}$	$\frac{\sqrt{1}}{\sqrt{2}}^{(15)}$	.											
M=5,4,3	13 2							$\sqrt{\frac{3}{2}}^{(23)}$	$1^{(34)}$	.	.	.	$\frac{\sqrt{3}}{\sqrt{2}}^{(13)}$	$\frac{1}{2}^{(24)}$	$\sqrt{\frac{3}{2}}^{(24)}$	$1^{(35)}$	.	.	$-\frac{1}{2}^{(14)}$	$-\frac{\sqrt{3}}{2}^{(14)}$	.	$\frac{1}{2}^{(25)}$	$\frac{\sqrt{3}}{2}^{(25)}$	.	.	$-\frac{1}{2}^{(15)}$	.	$-\frac{\sqrt{3}}{2}^{(15)}$	.	.					
	12 3							$\sqrt{\frac{1}{2}}^{(23)}$	.	$\sqrt{2}^{(12)}$	$1^{(34)}$	.	$\sqrt{\frac{1}{2}}^{(13)}$	$\sqrt{\frac{3}{2}}^{(24)}$	$-\frac{1}{2}^{(24)}$	.	.	$1^{(35)}$	$\frac{\sqrt{3}}{2}^{(14)}$	$-\frac{1}{2}^{(14)}$	.	$\frac{\sqrt{3}}{2}^{(25)}$	$-\frac{1}{2}^{(25)}$	.	.	$\frac{\sqrt{3}}{2}^{(15)}$	.	$-\frac{1}{2}^{(15)}$	.	.					
	11 4							.	.	.	$\sqrt{2}^{(12)}$	$1^{(45)}$	.	.	$\sqrt{2}^{(13)}$	.	.	.	.	.	$1^{(14)}$	.	.	.	.	.	.	.	$\frac{\sqrt{3}}{\sqrt{2}}^{(15)}$	$\frac{\sqrt{1}}{\sqrt{2}}^{(15)}$					
M=5,4,3,2	13 3												$1^{(12)}$	$\sqrt{\frac{3}{2}}^{(34)}$	$\sqrt{\frac{1}{2}}^{(34)}$	.	.	.	.	.	.	$\sqrt{\frac{3}{2}}^{(35)}$	$\sqrt{\frac{1}{2}}^{(35)}$	.	.	.	$-1^{(14)}$	.	.	.	.	.	.	$-1^{(15)}$	.
	14 2												.	$1^{(23)}$	.	$1^{(45)}$	.	.	$-1^{(13)}$	.	$\sqrt{\frac{3}{2}}^{(24)}$	.	.	.	$-\frac{\sqrt{3}}{\sqrt{2}}^{(14)}$	.	.	.	$\frac{1}{2}^{(25)}$	$\frac{\sqrt{3}}{2}^{(25)}$	.	$-\frac{1}{2}^{(15)}$	$-\frac{\sqrt{3}}{2}^{(15)}$	.	.
	22 3												$1^{(23)}$	.	.	.	$1^{(34)}$	.	$\frac{\sqrt{3}}{2}^{(24)}$	$-\frac{\sqrt{1}}{\sqrt{2}}^{(24)}$	.	.	.	$1^{(35)}$	.	$\frac{\sqrt{3}}{2}^{(25)}$	.	$-\frac{\sqrt{1}}{\sqrt{2}}^{(25)}$	.	.	.	.	.	.	.
	12 4												.	.	$1^{(23)}$	.	$\sqrt{2}^{(12)}$	$1^{(45)}$	.	$1^{(13)}$	$\sqrt{\frac{1}{2}}^{(24)}$	.	.	.	$\sqrt{\frac{1}{2}}^{(14)}$	.	.	.	$\frac{\sqrt{3}}{2}^{(25)}$	$-\frac{1}{2}^{(25)}$	.	$\frac{\sqrt{3}}{2}^{(15)}$	$-\frac{1}{2}^{(15)}$	.	.
	11 5												.	.	.	.	.	$\sqrt{2}^{(12)}$	.	.	.	.	$\sqrt{2}^{(13)}$	.	.	.	.	.	.	$\sqrt{2}^{(14)}$	.	.	.	.	$1^{(15)}$

(Eq. 24)

from Table III-d for V^3 into Eq.23 to give the rectangular matrix $\langle V^3 \cdot V^3 \rangle$ in Eq.25.

$$\langle V^3 \cdot V^3 \rangle = \frac{1}{10} \begin{array}{c|ccccc|} & \begin{smallmatrix}13\\3\end{smallmatrix} & \begin{smallmatrix}14\\2\end{smallmatrix} & \begin{smallmatrix}22\\3\end{smallmatrix} & \begin{smallmatrix}12\\4\end{smallmatrix} & \begin{smallmatrix}11\\5\end{smallmatrix} \\ \hline \begin{smallmatrix}13\\3\end{smallmatrix} & 32 & 2\sqrt{6} & -2\sqrt{6} & 2\sqrt{2} & -10 \\ \begin{smallmatrix}22\\3\end{smallmatrix} & -2\sqrt{6} & 0 & 46 & -4\sqrt{3} & 0 \\ \hline \end{array} \tag{25}$$

The application of this to the vectors $|I\rangle$ and $|II\rangle$ gives the 2x2 matrix h_{ij} (Eq.26a) and the desired interaction matrix (Eq26b). (At the same time we can find the eigenvalues of the ^2F, ^2G, and ^2H states, (Eq.26c) although these could have been found more quickly at higher levels.)

$$\boxed{\langle V^3 \cdot V^3 \rangle} \; \boxed{|I\rangle \;\; |II\rangle} = \begin{matrix} \frac{13}{70} & -\sqrt{\frac{6}{5}} \\ -\frac{27\sqrt{3}}{5\sqrt{35}} & \frac{54}{10\sqrt{5}} \end{matrix} \tag{26a}$$

(26b)

$$V^3 \cdot V^3 |I\rangle = \frac{13}{5}|I\rangle - \frac{2\sqrt{21}}{5}|II\rangle$$

$$V^3 \cdot V^3 |II\rangle = \frac{-2\sqrt{21}}{5}|I\rangle + \frac{21}{5}|II\rangle$$

(26c)

$$V^3 \cdot V^3 |^2F\rangle = \frac{24}{5}|^2F\rangle$$

$$V^3 \cdot V^3 |^2G\rangle = 4|^2G\rangle$$

$$V^3 \cdot V^3 |^2H\rangle = 3|^2H\rangle$$

It is interesting to find the eigenvectors of this operator $V^3 \cdot V^3$ since it is related through Eq.27 to the operators used by Racah (Recall Eq.9 of (I).) to label the repeated states.

$$V^3 \cdot V^3 = (M - P) / 2 - V^1 \cdot V^1 \tag{27}$$

The M and the $V^1 \cdot V^1$ are multiples of the unit matrix in the ($|I\rangle$, $|II\rangle$) representation, (M is the U_5 invariant, while $V^1 \cdot V^1$ is proportional to L^2) so eigenvectors $|\nu\rangle, |\nu'\rangle$ of $V^3 \cdot V^3$ in Eq.28 must also be eigenvectors of the pairing operator.

(28)

$$|\nu, \langle V^3 \cdot V^3 \rangle = \tfrac{27}{5} \rangle = (3\sqrt{2}\ \overset{13}{3} - 3\sqrt{3}\ \overset{14}{2} - 4\sqrt{3}\ \overset{22}{3} + 5\ \overset{12}{4} - 5\sqrt{2}\ \overset{11}{5}\ ) / 2\sqrt{42}$$

$$|\nu', \langle V^3 \cdot V^3 \rangle = \tfrac{7}{5} \rangle = (\sqrt{6} \qquad -3 \qquad 0 \qquad -\sqrt{3} \qquad \sqrt{6}\ ) / 2\sqrt{6}$$

Similar relations in Eq.29 involve $V^3 \cdot V^3$ and $V^5 \cdot V^5$ in the f sub-shell.

(29)

$$V^3 \cdot V^3 = (M - P) / 2 - G$$

$$V^3 \cdot V^3 + V^5 \cdot V^5 = (M - P) / 2 - V^1 \cdot V^1$$

It is interesting to see how the second state in Eq.28 can be derived directly in another way. First we write the two particle state of a pair (in Eq.30), namely a d^2 orbital scalar with spin 0.

$$
\begin{aligned}
|\text{pair}\rangle &= |d^2\ {}^1S\rangle \qquad\qquad (30)\\
&= \frac{(|1\rangle|5\rangle + |5\rangle|1\rangle - |2\rangle|4\rangle - |4\rangle|2\rangle + |3\rangle|3\rangle)}{\sqrt{5}}\frac{(|\uparrow\rangle|\downarrow\rangle - |\downarrow\rangle|\uparrow\rangle)}{\sqrt{2}}\\
&= \left(\sqrt{2}\left|15\begin{smallmatrix}\uparrow\\ \downarrow\end{smallmatrix}\right\rangle - \sqrt{2}\left|24\begin{smallmatrix}\uparrow\\ \downarrow\end{smallmatrix}\right\rangle + \left|33\begin{smallmatrix}\uparrow\\ \downarrow\end{smallmatrix}\right\rangle\right)/\sqrt{5}
\end{aligned}
$$

The last line in Eq.30 above uses tableau notation, and the assembly formula (Fig.10) converts this to a sum of Slater Determinants in Eq.31 below.

$$
|\text{pair}\rangle = \left(\begin{matrix}1\uparrow\\5\downarrow\end{matrix} - \begin{matrix}1\downarrow\\5\uparrow\end{matrix} - \begin{matrix}2\uparrow\\4\downarrow\end{matrix} + \begin{matrix}2\downarrow\\4\uparrow\end{matrix} + \begin{matrix}3\uparrow\\3\downarrow\end{matrix}\right)/\sqrt{5} \qquad (31)
$$

Now the state we want turns out to be this pair plus one more d particle, all properly antisymmetrized. Such a state is constructed in Eq.32 by simply attaching a $1\uparrow$ to Eq.31. In the first line of Eq.32 we drop the zero state written in brackets. The second line of Eq.32 comes from applying the assembly formula again and gives a result which is the same as the second state in Eq.28, except for an overall phase.

$$
\begin{aligned}
|\text{pair} + 1\uparrow\rangle &= \left(\begin{bmatrix}1\uparrow\\1\uparrow\\5\downarrow\end{bmatrix} - \begin{matrix}1\uparrow\\1\downarrow\\5\uparrow\end{matrix} - \begin{matrix}1\uparrow\\2\uparrow\\4\downarrow\end{matrix} + \begin{matrix}1\uparrow\\2\downarrow\\4\uparrow\end{matrix} + \begin{matrix}1\uparrow\\3\uparrow\\3\downarrow\end{matrix}\right)/2 \qquad (32)\\
&= \left(-\left|\begin{smallmatrix}11\\5\end{smallmatrix}\ \begin{smallmatrix}\uparrow\uparrow\\ \downarrow\end{smallmatrix}\right\rangle + \sqrt{\tfrac{2}{3}}\left|\begin{smallmatrix}14\\2\end{smallmatrix}\ \begin{smallmatrix}\uparrow\uparrow\\ \downarrow\end{smallmatrix}\right\rangle + \left(\sqrt{\tfrac{1}{2}}\left|\begin{smallmatrix}12\\4\end{smallmatrix}\ \begin{smallmatrix}\uparrow\uparrow\\ \downarrow\end{smallmatrix}\right\rangle + \sqrt{\tfrac{1}{6}}\left|\begin{smallmatrix}14\\2\end{smallmatrix}\ \begin{smallmatrix}\uparrow\uparrow\\ \downarrow\end{smallmatrix}\right\rangle\right) - \left|\begin{smallmatrix}13\\3\end{smallmatrix}\ \begin{smallmatrix}\uparrow\uparrow\\ \downarrow\end{smallmatrix}\right\rangle\right)/2\\
&= \left(-\sqrt{6}\begin{matrix}11\\5\end{matrix} + 3\begin{matrix}14\\2\end{matrix} + \sqrt{3}\begin{matrix}12\\4\end{matrix} - \sqrt{6}\begin{matrix}13\\3\end{matrix}\right)\left(\begin{matrix}\uparrow\uparrow\\ \downarrow\end{matrix}\right)/2\sqrt{6}
\end{aligned}
$$

The preceeding indicates some ways to relate tableaus and Racah eigenvectors, and a more complete analysis of this will be given in future works [29]. Meanwhile, we expect that most applications will just need the eigenvectors of the electrostatic interaction to carry out further calculations involving spin-orbit, radiation, and crystal fields, so we review now the relations between the $V^k \cdot V^k$ and the electrostatic operator.

The subtraction of the "self energy" from $V^k \cdot V^k$ defines a $V(k \cdot k)$ operator [30] (eq. 33) which appears in the Coulomb energy formula of Eqs. 34.

$$V(k\cdot k) = \frac{1}{2}\sum_{\alpha\neq\beta}\sum_{q}(-1)^q\, v_q^k\,(\text{electron }\alpha)\; v_q^k\,(\text{electron }\beta) \tag{33}$$

$$= \frac{1}{2}\, V^k \cdot V^k - \frac{1}{2}\sum_m E_{mm}\left(\frac{2k+1}{2\ell+1}\right)$$

General derivations of these formulas are given in the following section.

$$\left\langle \ell^n \gamma' L \right| \frac{1}{2}\sum_{\alpha\neq\beta}\frac{1}{|r_\alpha - r_\beta|}\left| \ell^n \gamma\, L\right\rangle \tag{34}$$

$$= \sum_k \left\langle \ell^n \gamma' L\right| \frac{(2\ell+1)^2}{(2k+1)}\begin{pmatrix}\ell & k & \ell\\ 0 & 0 & 0\end{pmatrix}^2 F^k\, V(k\cdot k)\left|\ell^n \gamma\, L\right\rangle$$

for $\ell=1$ $= 3F^0\langle V(0\cdot 0)\rangle + \frac{6}{25}F^2\langle V(2\cdot 2)\rangle$

for $\ell=2$ $= 5F^0\langle V(0\cdot 0)\rangle + \frac{2}{7}F^2\langle V(2\cdot 2)\rangle + \frac{10}{63}F^4\langle V(4\cdot 4)\rangle$

for $\ell=3$ $= 7F^0\langle V(0\cdot 0)\rangle + \frac{28}{75}F^2\langle V(2\cdot 2)\rangle + \frac{14}{99}F^4\langle V(4\cdot 4)\rangle + \frac{700}{5577}F^6\langle V(6\cdot 6)\rangle$

The standard Slater radial integral of Eq. 35 is used.

$$F^k = \int r_1^2\, dr_1 \int r_2^2\, dr_2\, \frac{r_<^k}{r_>^{k+1}}\, R_\ell^2(r_1)\, R_\ell^2(r_2) \tag{35}$$

If the repeated orbital states of each multiplet are defined by the electrostatic operator, or by the preceeding orthogonalization procedure, or even by artificial labeling sets such as the Racah operators, then no further ambiguity can arise when spin is considered. Hence, the assembly formula and related algorithms for treating spin and orbit operators described in (I) can be applied in the same manner as was given there.

B. MIXED ORBITAL CONFIGURATIONS $(\ell_1)^{n_1}\ (\ell_2)^{n_2}\ \ldots$

The following discussion of the unitary basis for several sub-shells is meant to explain additional calculational procedures, and to show further structure of the bases and operators.

If a single electron can occupy several orbital subshells $n\ell_1$, $n'\ell_2$, ... there will be $m = (2\ell_1+1) + (2\ell_2+1) + \ldots$ orbital states for it. Also there will be a set of m^2 (1-body) operators for these states, which will generate the group $U_m = U_{(2\ell_1+1)+(2\ell_2+1)+\ldots}$. (We shall treat an example in which electrons can occupy a d or a p sub-shell, and this will involve the group U_8.) These operators can be taken to be the m^2 elementary operators E_{ij} or else the m^2 tensor operators defined in Eq.36.

$$V^k_q(\ell_1\ell_2) = \sum_{m_1m_2} \left|\begin{smallmatrix}\ell_1\\ m_1\end{smallmatrix}\right\rangle\left\langle\begin{smallmatrix}\ell_1\\ m_1\end{smallmatrix}\right| V^k_q(\ell_1\ell_2)\left|\begin{smallmatrix}\ell_2\\ m_2\end{smallmatrix}\right\rangle\left\langle\begin{smallmatrix}\ell_2\\ m_2\end{smallmatrix}\right| \tag{36a}$$

$$= \sum_{m_1m_2} \left\langle\begin{smallmatrix}\ell_1\\ m_1\end{smallmatrix}\right| V^k_q(\ell_1\ell_2)\left|\begin{smallmatrix}\ell_2\\ m_2\end{smallmatrix}\right\rangle E_{(\ell_1m_1)(\ell_2m_2)} \tag{36b}$$

$$\left\langle\begin{smallmatrix}\ell_1\\ m_1\end{smallmatrix}\right| V^k_q(\ell_1\ell_2)\left|\begin{smallmatrix}\ell_2\\ m_2\end{smallmatrix}\right\rangle = \frac{(-1)^{\ell_1-m_1}}{\sqrt{2k+1}}\begin{pmatrix}\ell_1 & k & \ell_2\\ -m_1 & q & m_2\end{pmatrix} = C^{k\ \ell_2\ell_1}_{q\ m_2m_1}(-1)^{\ell_1+\ell_2+k}\left[(2k+1)/(2\ell_1+1)\right]^{\frac{1}{2}}$$

For $\ell_1 = \ell_2$ Eq.36 reduces to Eq.1 and gives the matrices in Tables II-III. For non-zero "shift" $\Delta = \ell_1 - \ell_2$ the matrices of the operators will be rectangular as shown in Table IV, or in the examples given by Eq.37 below.

$$\sqrt{10}\, V^1_1(pd) = \begin{bmatrix} \cdot & \cdot & 1 & \cdot & \cdot \\ \cdot & \cdot & \cdot & \sqrt{3} & \cdot \\ \cdot & \cdot & \cdot & \cdot & \sqrt{6} \end{bmatrix} = E_{63} + \sqrt{3}\, E_{74} + \sqrt{6}\, E_{85} \qquad \sqrt{10}\, V^1_{-1}(dp) = \begin{bmatrix} \cdot & \cdot & \cdot \\ \cdot & \cdot & \cdot \\ 1 & \cdot & \cdot \\ \cdot & \sqrt{3} & \cdot \\ \cdot & \cdot & \sqrt{6} \end{bmatrix} = E_{36} + \sqrt{3}E_{47} + \sqrt{6}E_{58} \tag{37}$$

The numbering for E_{ij} reflects the choice of numbers 1-5 for d states and 6-8 for the p states. The tables exhibit the $V^k_q(\ell_1 \ell_2)$ matrices for $\ell_1 - \ell_2 \equiv \Delta > 0$, and the transpose is found using the symmetry relation in Eq.38.

$$V^k_q(\ell_2 \ell_1) = (-1)^{\Delta + q}\, \widetilde{V^k_{-q}}(\ell_1 \ell_2) \tag{38}$$

To see the implied physical distinction between operators with $\Delta = 0$ on one hand, and those with $\Delta \neq 0$ on the other, we may compare the two types of vector (k=1) operators which we will be using shortly. The $\Delta = 0$ operators $V^1_q(pp)$ or $V^1_q(dd)$ correspond to components of the angular momentum operator L_q or any other polar vector operators like those of magnetic dipole. The $\Delta = \pm 1$ operators $V^1_q(pd)$ or $V^1_q(dp)$ correspond to the electric dipole or any other axial vector operator. The former conserves parity while the latter changes it.

For examples of mixed configurations we shall treat some states arising from $(d)^n(p)^{n'}$. The angular momentum lowering operator in this basis is given by Eq.39. (Eq.4 and Eq.17d of (I) were combined using the same numbering as Eq.37 above.)

$$L_- = \sqrt{20}\, V^1_{-1}(dd) + 2V^1_{-1}(pp) \tag{39}$$

$$= 2E_{21} + \sqrt{6}E_{32} + \sqrt{6}E_{43} + 2E_{54} + \sqrt{2}E_{76} + \sqrt{2}E_{87}$$

Now for example, the orbital parts of the doublet ($S=\frac{1}{2}$) terms, which three particles can make when occupying d and p sub-shells, belong to the (2100...0) representation of U_8. In Eq. 40 the dimension formula (Fig.6-b) gives the total number (168) of orbital states.

$$\frac{\begin{matrix}89\\7\end{matrix}}{\begin{matrix}31\\1\end{matrix}} = 168 \tag{40}$$

Fortunately we can deal individually with smaller subsets of such manifolds. First of all, we have already treated in Sec. 1A some of the 40 orbital doublets of the pure configuration $(d^3\ p^0)$. Only after this comes the remaining 128 "excited" doublet states arising from configurations (d^2p), $(d\ p^2)$, and (d^0p^3). The unitary calculus can give all possible matrix elements between all these configurations since they all belong to one representation.

The states and terms of a mixed configuration like (d^2p) are found by systematically lowering "highest tableaus" just as was done for the pure configuration $(d)^3$ in Sec.1A. The lowering is done with all the operators $E_{n\ n-1}$ except now we exclude any inter-shell operators such as the E_{65} in our example. Consequently several "highest tableau" (These are all lexical tableau made from the configuration that cannot be raised by any intra-shell operator $E_{n\ n+1}$.) may exist for a given configuration, and for each there arise a separate family of states and terms. For (d^2p) there are two highest tableaus $\begin{smallmatrix}11\\6\end{smallmatrix}$ and $\begin{smallmatrix}16\\2\end{smallmatrix}$ which give the states and matrices shown partly in Eq.41 and 42 respectively. L_- in Eq.39 gives the highest terms which are written below each of these matrices.

For configuration (dp^2) there is just one highest tableau, and the lowering of this gives the states shown in Eq.43. The orbital states labeled in Eq.41-43 plus the 48 states coming from $(d)^3$ and $(p)^3$ account for the 168 states predicted by Eq.40.

Before proceeding, we shall see how this manifold would appear if the p states 6,7, and 8 were ordered <u>before</u> the d states 1, 2,...5. For example p^2d gives rise to the same number of states as came from dp^2 in Eq.43, but they come in two separate families as shown in Eq.44 and 45. Clearly a linear transformation exists between the two choices of ordering, and the assembly formula can give this.

$$\langle L_- \ E_{mn} \rangle = \tag{44}$$

G	F				D	D					P						S
$\begin{smallmatrix}66\\1\end{smallmatrix}$	$\begin{smallmatrix}67\\1\end{smallmatrix}$	$\begin{smallmatrix}66\\2\end{smallmatrix}$	$\begin{smallmatrix}77\\1\end{smallmatrix}$	$\begin{smallmatrix}68\\1\end{smallmatrix}$	$\begin{smallmatrix}67\\2\end{smallmatrix}$	$\begin{smallmatrix}66\\3\end{smallmatrix}$	$\begin{smallmatrix}78\\1\end{smallmatrix}$	$\begin{smallmatrix}77\\2\end{smallmatrix}$	$\begin{smallmatrix}68\\2\end{smallmatrix}$	$\begin{smallmatrix}67\\3\end{smallmatrix}$	$\begin{smallmatrix}66\\4\end{smallmatrix}$	$\begin{smallmatrix}88\\1\end{smallmatrix}$	$\begin{smallmatrix}78\\2\end{smallmatrix}$	$\begin{smallmatrix}77\\3\end{smallmatrix}$	$\begin{smallmatrix}68\\3\end{smallmatrix}$	$\begin{smallmatrix}67\\4\end{smallmatrix}$	$\begin{smallmatrix}66\\5\end{smallmatrix}$
	$\sqrt{2}$	1															
2																	
2																	

$$|(p^2\ {}^1L)d\ \ {}^2G,4\rangle = \left|\begin{smallmatrix}66\\1\end{smallmatrix}\right\rangle$$

$$|(p^2\ {}^1L)d\ \ {}^2F,3\rangle = \left(\left|\begin{smallmatrix}67\\1\end{smallmatrix}\right\rangle - \left|\begin{smallmatrix}66\\2\end{smallmatrix}\right\rangle\right)/\sqrt{2}$$

	H	G				F	F					D						P	P								
	11	12	11	22	13	12	11	23	22	14	13	12	33	24	23	22	15	14	13	34	33	25	24	23	15	14	... = $\{d^2\,{}^1Lp\}$
	6	6	7	6	6	7	8	6	7	6	7	8	6	6	7	8	6	7	8	6	7	6	7	8	7	8	
	.	$\sqrt{2}^{12}$	1^{67}	.	$\sqrt{2}^{13}$	.	1^{68}	.	.	$\sqrt{2}^{14}$	.	.	.	.	.	.	$\sqrt{2}^{15}$	.	.	.	.	.	.	.	.	.	
$2\sqrt{2}$		.	.	$\sqrt{2}^{12}$	1^{23}	1^{67}	.	1^{13}	.	1^{24}	.	1^{68}	.	1^{14}	.	.	1^{25}	.	.	.	.	1^{15}	.	.	.	.	
$\sqrt{2}$		.	.	.	.	$\sqrt{2}^{12}$	1^{78}	.	.	.	$\sqrt{2}^{13}$	.	.	.	.	.	.	$\sqrt{2}^{14}$	.	.	.	.	.	.	$\sqrt{2}^{15}$	.	$= \langle L_- \, E_{mn} \rangle$

(41a)

$$^2H5 = {11 \atop 6}$$

$$^2H4 = \left(2\,{12 \atop 6} + {11 \atop 7}\right) / \sqrt{5}$$

$$^2G4 = \left({12 \atop 6} - 2\,{11 \atop 7}\right) / \sqrt{5}$$

(41b)

	G		F			D	D					P					S	
	16	16	17	26	16	17	18	26	27	16	17	18	36	26	27	28	17	18	... = $\{d^2\,{}^3L\,p\}$
	2	3	2	3	4	3	2	4	3	5	4	3	4	5	4	3	5	4	
	.	1^{23}	1^{67}	-1^{13}	1^{24}	.	1^{68}	-1^{14}	.	1^{25}	.	.	.	-1^{15}	.	.	.	.	
$\sqrt{6}$		.	.	1^{12}	1^{34}	1^{67}	.	.	.	1^{35}	.	1^{68}	-1^{14}	.	.	.	.	.	
$\sqrt{2}$		.	.	.	.	1^{23}	1^{78}	.	-1^{13}	.	1^{24}	.	.	.	-1^{14}	.	1^{25}	.	$= \langle L_- \, E_{mn} \rangle$

(42a)

$$^2G4 = {16 \atop 2}$$

$$^2G3 = \left(\sqrt{3}\,{16 \atop 3} + {17 \atop 2}\right) / 2$$

(42b)

G		F	F				D	D	D							P	P									S	
16	26	17	16	36	27	26	18	17	16	46	37	36	28	27	26	18	17	56	47	46	38	37	36	28	27	18	... = $\{dp^2\}$
6	6	6	7	6	6	7	6	7	8	6	6	7	6	7	8	7	8	6	6	7	6	7	8	7	8	8	

(43)

(45)

$$\left\langle L_{-} \cdot\cdot E_{mn} \right\rangle = \begin{array}{c|cc|ccc|ccc} & F & & D & & & P & & \\ 61 & 61 & 62 & 71 & 62 & 63 & 72 & 63 & 64 \\ 7 & 8 & 7 & 8 & 8 & 7 & 8 & 8 & 7 \\ \hline & 1 & 1 & & & & & & \\ 2 & & & & & & & & \\ 2 & & & & & & & & \end{array} \cdots \qquad \left|(p^2\ {}^3L)d\ {}^2F,3\right\rangle = \left|\begin{smallmatrix}61\\7\end{smallmatrix}\right\rangle$$

Some results of this transformation are given in Eq.46. It is seen that separate and unambiguous choices are made for two ^2F terms that came together in the dp^2 manifold of Eq.43.

$$\left|(p^2\ {}^1L)d\ {}^2G,4\right\rangle = \left|\begin{smallmatrix}66\\1\end{smallmatrix}\right\rangle = -\left|\begin{smallmatrix}16\\6\end{smallmatrix}\right\rangle \tag{46}$$

$$\left|(p^2\ {}^3L)d\ {}^2F,3\right\rangle = \left|\begin{smallmatrix}61\\7\end{smallmatrix}\right\rangle = \left(-\left|\begin{smallmatrix}17\\6\end{smallmatrix}\right\rangle + \sqrt{3}\left|\begin{smallmatrix}16\\7\end{smallmatrix}\right\rangle\right) / 2$$

$$\left|(p^2\ {}^1L)d\ {}^2F,3\right\rangle = \left(\left|\begin{smallmatrix}67\\1\end{smallmatrix}\right\rangle - \left|\begin{smallmatrix}66\\2\end{smallmatrix}\right\rangle\right) / \sqrt{2} = \left(2\left|\begin{smallmatrix}26\\6\end{smallmatrix}\right\rangle - \sqrt{3}\left|\begin{smallmatrix}17\\6\end{smallmatrix}\right\rangle - \left|\begin{smallmatrix}16\\7\end{smallmatrix}\right\rangle\right) / \sqrt{8}$$

These choices turn out to be eigenstates in the absence of exchange interaction, since their parentage is traceable to simple products of lesser configurations.

For machine calculation the value of tracing parentage is probably negligible since the electrostatic interaction will mix up the results anyway. The important thing is that all operators on tableau states can be evaluated for any choice of ordering using the same formulas that work for pure configurations. For example, to evaluate the relative dipole matrix elements for various possible $^2G \leftrightarrow {}^2H$ transitions indicated in Fig.11, we need only to compute the relevant elementary operators as is done below in Eqs.47.

(47)

$$\langle (d^2\ ^1Gp)\ ^2G,4 |\ V^1_{-1}(pd) |\ d^3\ ^2H,5 \rangle = \frac{\langle {12 \atop 6} | - 2 \langle {11 \atop 7} |}{\sqrt{5}} \ \frac{\sqrt{6}E_{61} + \sqrt{3}E_{72} + E_{83}}{\sqrt{10}} \left| {11 \atop 2} \right\rangle$$

$$= (\sqrt{6} \langle {12 \atop 6} | E_{61} | {11 \atop 2} \rangle - 2\sqrt{3} \langle {11 \atop 7} | E_{72} | {11 \atop 2} \rangle)/\sqrt{50}$$

$$= (-\sqrt{6}\sqrt{\tfrac{1}{2}} \qquad - 2\sqrt{3} \ )/\sqrt{50}$$

$$= -3\sqrt{3/50}$$

$$\langle d^3\ ^2G,4 |\ V^1_{-1}(dp) | (d^2\ ^1G\ p)\ ^2H,5 \rangle = -2 \langle {11 \atop 3} | E_{36} | {11 \atop 6} \rangle /10$$

$$= -2\ /\ 10$$

$$\langle (d^2\ ^3F\ p)\ ^2G,4 |\ V^1_{-1}(pd) |\ d^3\ ^2H \rangle = \langle {16 \atop 2} | \frac{\sqrt{6}E_{61}}{\sqrt{10}} | {11 \atop 2} \rangle$$

$$= \frac{3}{\sqrt{10}}$$

The actual transition intensities will depend mainly on quantities above and the mixing of repeated states effected by the electrostatic interaction. A general discussion of the unitary analysis of the electrostatic interaction is given now.

We use the addition theorem formula (Eq.48a) in the electrostatic operator for each pair of electrons α and β, where the C^k_q (Eq.48b) are the standard multipole functions.

(48a)

$$\frac{1}{|r_{\alpha\beta}|} = \sum_k \sum_q (-1)^q \frac{r_<^k}{r_>^{k+1}} \ C^k_{-q}(\alpha)\ C^k_q(\beta)$$

$$C_q^k(\alpha) = \sqrt{\frac{4\pi}{2k+1}}\; Y_q^k(\theta_\alpha, \phi_\alpha) \tag{48b}$$

Inserting completeness sums over all sub-shells of interest will convert Eq.48 into an expression for the effective electrostatic operator in these sub-shells. The result is the combination in Eq.49 of elementary operators $e_{ij}(\gamma)$ for the electrons α and β.

(49a)

$$\frac{1}{|r_{\alpha\beta}|} \rightarrow \sum_{\substack{l_1 l_2 l_1' l_2' \\ m_1 m_2 m_1' m_2'}} \left|\begin{matrix} l_1' l_2' \\ m_1' m_2' \end{matrix}\right\rangle \left\langle\begin{matrix} l_1' l_2' \\ m_1' m_2' \end{matrix}\right| \frac{1}{|r_{\alpha\beta}|} \left|\begin{matrix} l_1 l_2 \\ m_1 m_2 \end{matrix}\right\rangle \left\langle\begin{matrix} l_1 l_2 \\ m_1 m_2 \end{matrix}\right|$$

$$= \sum_{\substack{l_1 l_2 l_1' l_2' \\ m_1 m_2 m_1' m_2'}} \underset{\substack{l_1' l_1 \\ m_1' m_1}}{e}(\alpha) \; \underset{\substack{l_2' l_2 \\ m_2' m_2}}{e}(\beta) \; \sum_k F^k(l_1' l_2' l_1 l_2) \cdot$$

$$\left[\sum_q (-1)^q \left\langle\begin{matrix} l_1' \\ m_1' \end{matrix}\right| C_{-q}^k(\alpha) \left|\begin{matrix} l_1 \\ m_1 \end{matrix}\right\rangle \left\langle\begin{matrix} l_2' \\ m_2' \end{matrix}\right| C_q^k(\beta) \left|\begin{matrix} l_2 \\ m_2 \end{matrix}\right\rangle\right]$$

(49b)

$$F^k(l_1' l_2' l_1 l_2) = \int r_1^2\, dr_1 \int r_2^2\, dr_2\; R_{l_1'}(r_1)\, R_{l_2'}(r_2)\, \frac{r_<^k}{r_>^{k+1}}\, R_{l_1}(r_1)\, R_{l_2}(r_2)$$

The algebraic expression [31] for matrix elements of C_q^k gives the following form (Eq.50) for this effective operator. The quantities $\left(\begin{smallmatrix} & k & \\ i & & j \end{smallmatrix}\right)$ in Eq.50a are the elementary coefficients of the tensors V^k given in Tables II-IV or by Eq.36b. The modified radial integral A^k in Eq.50c has the parity selection rules $(-1)^{l_1'+k+l_1} = 1 = (-1)^{l_2'+k+l_2}$ built into it. These were used to obtain the phase factor $(-1)^\Delta = (-1)^{l_1-l_1'} = (-1)^{l_2-l_2'}$ in Eq.50a.

(50a)

$$\frac{1}{|r_{\alpha\beta}|} \rightarrow \sum_{\substack{l_1' l_2' l_1 l_2 \\ m_1 m_2}} \sum_k A^k(l_1' l_2' l_1 l_2) \sum_{q=-k}^{q=k} (-1)^{q+\Delta} \binom{k}{1'\,1} e_{1'1}(\alpha) \binom{k}{2'\,2} e_{2'2}^{(\beta)}$$

$$\binom{k}{1'\,1} = C^{k\ l_1\ l_1'}_{-q\ m_1\ m_1-q} \sqrt{(2k+1)/(2l_1'+1)} \qquad (50b)$$

$$1' \rightarrow {}^{l_1'}_{m_1'} \qquad 1 \rightarrow {}^{l_1}_{m_1}$$

$$\binom{k}{2'\,2} = C^{k\ l_2\ l_2'}_{q\ m_2\ m_2+q} \sqrt{(2k+1)/(2l_2'+1)}$$

$$2' \rightarrow {}^{l_2'}_{m_2'} \qquad 2 \rightarrow {}^{l_2}_{m_2}$$

(50c)

$$A^k(l_1' l_2' l_1 l_2) = F^k(l_1' l_2' l_1 l_2) \begin{pmatrix} l_1' & k & l_1 \\ 0 & 0 & 0 \end{pmatrix} \begin{pmatrix} l_2' & k & l_2 \\ 0 & 0 & 0 \end{pmatrix} \sqrt{(2l_1+1)(2l_2+1)(2l_1'+1)(2l_2'+1)/(2k+1)}$$

The total electrostatic operator is a combination (Eq.51) of the total elementary operators $E=\sum_\alpha e(\alpha)$. The sum over $\alpha=\beta$ has been subtracted from the sum over all α and β. In the subtracted term the basic property (Eq.51b) of the single particle elementary operator was used.

(51a)

$$\frac{1}{2}\sum_{\alpha\neq\beta} \frac{1}{|r_{\alpha\beta}|} \rightarrow \frac{1}{2} \sum_{l_1' l_2' l_1 l_2} \sum_k A^k(l_1' l_2' l_1 l_2) \Big[\sum_{\substack{q \\ m_1 m_2}} (-1)^{q+\Delta} \binom{k}{1'\,1} E_{1'1} \binom{k}{2'\,2} E_{2'2} - \sum_{\substack{q \\ m_1 m_2}} (-1)^{q+\Delta} \binom{k}{1'\,1} \binom{k}{2'\,2} \delta_{2'1} E_{1'2} \Big]$$

$$e_{ij}(\alpha)\, e_{km}(\alpha) = \delta_{jk}\, e_{im}(\alpha) \qquad (51b)$$

The coefficients in the first term give immediately a simple expression involving the total V tensors. (The transpose relation in Eq.38 was used.) The orthonormality of these same coefficients reduce the subtracted term to a sum over the number operator $n(\ell) = \sum_m E_{\ell\ell}^{\;mm}$ for each sub-shell.

$$\frac{1}{2} \sum_{\alpha \neq \beta} \frac{1}{|r_{\alpha\beta}|} \longrightarrow \tag{52}$$

$$\frac{1}{2} \sum_{\ell_1' \ell_2' \ell_1 \ell_2} \sum_k A^k(\ell_1' \ell_2' \ell_1 \ell_2) \sum_q \tilde{V}_q^k(\ell_1 \ell_1') \, V_q^k(\ell_2' \ell_2)$$

$$-\frac{1}{2} \sum_{\ell_1 \ell_2} \sum_k A^k(\ell_1 \ell_2 \ell_2 \ell_1) \frac{2k+1}{2\ell_1+1} \sum_{m_1} E_{\ell_1 \ell_1}^{\;m_1 m_1}$$

Solving completely the electrostatic operator for $(d^n \; p^{n'})$ configurations would require the radial integrals A^k(dd dd) for k=0,2,4; A^k(dd pp) fpr k=1,3; A^k(pd pd) for k=0,2; A^k(dp pd) for k=1,3; A^k(pp pp) for k=0,2. All others are related to the above (A(ab cd)= A(ba dc)= A(cd ab)= A(dc ba)) or are zero (A(dd pd)=0) from parity considerations. Various approximate models can give useful results while ignoring certain of these quantities or treating them as undetermined parameters.

In any case, the angular coefficients of each radial integral require the matrix elements shown in Eq.53 below.

Coefficient of $A(\ell\ell\,\ell\ell)$ = (53)

$$\tfrac{1}{2}\Big\langle V_{o}(\ell\ell)V_{o}(\ell\ell)+\sum_{q=1}\big(\tilde{V}_{q}(\ell\ell)V_{q}(\ell\ell)+V_{q}(\ell\ell)\tilde{V}_{q}(\ell\ell)\big)\Big\rangle - \frac{n(\ell)}{2}\;\frac{2k+1}{2\ell+1}$$

Coefficient of $A(pd\;pd)$ =

$$\Big\langle V_{o}(pp)V_{o}(dd)+\sum_{q=1}\big(\tilde{V}_{q}(pp)V_{q}(dd)+V_{q}(pp)\tilde{V}_{q}(dd)\big)\Big\rangle$$

Coefficient of $A(dp\;pd)$ =

$$\tfrac{1}{2}\Big\langle \tilde{V}_{o}(pd)V_{o}(pd)+\tilde{V}_{o}(dp)V_{o}(dp)+\sum_{q=1}\big(\tilde{V}_{q}(pd)V_{q}(pd)+\tilde{V}_{q}(dp)V_{q}(dp)+V_{q}(pd)\tilde{V}_{q}(pd)+V_{q}(dp)\tilde{V}_{q}(dp)\big)\Big\rangle$$

$$-\;\frac{2k+1}{2}\left(\frac{n(p)}{3}+\frac{n(d)}{5}\right)$$

Coefficient of $A(dd\;pp)$ =

$$\tfrac{1}{2}\Big\langle \tilde{V}_{o}(pd)V_{o}(dp)+\tilde{V}_{o}(dp)V_{o}(pd)+\sum_{q=1}\big(\tilde{V}_{q}(pd)V_{q}(dp)+\tilde{V}_{q}(dp)V_{q}(pd)+V_{q}(pd)\tilde{V}_{q}(dp)+V_{q}(dp)\tilde{V}_{q}(pd)\big)\Big\rangle$$

The Coulomb matrix elements for the first dipole allowed excited terms in Fig.11 are given by the formulas in Eq.54 in terms of the elementary coefficients (ij) from Table III(d.p) or IV(d,p). These formulas give the components for general values of multipolarity k, including those that are zero for the Coulomb operator.

(54)

	$2A^k(dp\ pd)$	$A^k(pd\ pd)$	$2A^k(dd\ dd)$	$2A^k(pp\ pp)$
$\langle {}^{11}_{6} \vert \frac{1}{r} \vert {}^{11}_{6} \rangle =$	$(16)^2+2(17)^2+2(18)^2$ $+(26)^2$ $+(36)^2$ $+(46)^2$ $+(56)^2$ $-(2k+1)11/15$	$+2(11)(66)$	$+(2(11))^2$ $+2((12)^2+(13)^2+(14)^2+(15)^2)$ $-(2k+1)2/5$	$+(66)^2$ $+(67)^2+(68)^2$ $-(2k+1)1/3$
$\langle {}^{12}_{6} \vert \frac{1}{r} \vert {}^{12}_{6} \rangle =$	$(16)^2+(17)^2+(18)^2$ $+(26)^2+(27)^2+(28)^2$ $+(36)^2$ $+(46)^2$ $+(56)^2$ $-(2k+1)11/15$	$+(11)(66)$ $+(22)(66)$	$+((11)+(22))^2$ $+4((12)^2+(13)^2+(14)^2+(15)^2$ $+(23)^2+(24)^2+(25)^2$ $-(2k+1)2/5$	$+(66)^2$ $+(67)^2+(68)^2$ $-(2k+1)1/3$
$\langle {}^{12}_{6} \vert \frac{1}{r} \vert {}^{11}_{7} \rangle =$	$-\sqrt{2}(16)(27)$	$+\sqrt{2}(12)(67)$		
$\langle {}^{12}_{6} \vert \frac{1}{r} \vert {}^{16}_{2} \rangle =$	$\sqrt{3}((26)^2-(16)^2)$			
$\langle {}^{11}_{7} \vert \frac{1}{r} \vert {}^{11}_{7} \rangle =$	$2(16)^2+(17)^2+2(18)^2$ $+(27)^2$ $+(37)^2$ $+(47)^2$ $+(57)^2$ $-(2k+1)11/15$	$+2(11)(77)$	$+(2(11))^2$ $+2((12)^2+(13)^2+(14)^2+(15)^2)$ $-(2k+1)2/5$	$+(77)^2$ $+(67)^2+(78)^2$ $-(2k+1)1/3$
$\langle {}^{11}_{7} \vert \frac{1}{r} \vert {}^{16}_{2} \rangle =$	$\sqrt{6}(16)(27)$			
$\langle {}^{16}_{2} \vert \frac{1}{r} \vert {}^{16}_{2} \rangle =$	$3(16)^2+(17)^2+(18)^2$ $+3(26)^2+(27)^2+(28)^2$ $+(36)^2$ $+(46)^2$ $+(56)^2$ $-(2k+1)11/15$	$+(11)(66)$ $+(22)(66)$	$+((11)+(22))^2$ $+(13)^2+(14)^2+(15)^2$ $+(23)^2+(24)^2+(25)^2$ $-(2k+1)2/5$	$+(66)^2$ $+(67)^2+(68)^2$ $-(2k+1)1/3$

The coefficients of the pure integrals A(dd dd) and A(pp pp) have the same structure as they did for pure configurations. The coefficients of the exchange integrals involve diagonal terms such as Eq.55a, and off-diagonal terms such as Eq.55b. The commutation relations and diagonality of E_{jj} simplifies the calculations.

$$(ij)^2 \langle T | E_{ij}E_{ji} + E_{ji}E_{ij} | T\rangle = (ij)^2 \langle T | 2E_{ij}E_{ji} + E_{jj} - E_{ii} | T\rangle \tag{55a}$$

$$(ij)(km)\langle T' | E_{ij}E_{km} + E_{km}E_{ij} | T\rangle = 2(ij)(km)\langle T' | E_{ij}E_{km} | T\rangle \tag{55b}$$

Inspection of each tableau quickly determines which of these terms will contribute to each matrix element.

The total interaction Hamiltonian resulting from the exchange part of Eq.54 is given by Eq.56. The quantities H , G , and G$'$ in Eq.56 would include the direct $A(\ell\ell\,\ell\ell)$ terms.

(56)

$\langle \frac{1}{|r|} \rangle =$

$\vert{}^2H\rangle = \left\vert\begin{smallmatrix}11\\6\end{smallmatrix}\right\rangle$	$\vert{}^2G\rangle = \dfrac{\left\vert\begin{smallmatrix}12\\6\end{smallmatrix}\right\rangle - 2\left\vert\begin{smallmatrix}11\\7\end{smallmatrix}\right\rangle}{\sqrt{5}}$	$\vert{}^2G'\rangle = \left\vert\begin{smallmatrix}16\\2\end{smallmatrix}\right\rangle$
$-\frac{3}{5}A^1(\text{dp pd}) - \frac{1}{15}A^3(\text{dp pd})$ $\frac{2}{\sqrt{21}}A^2(\text{pd pd}) + H$		
	$\frac{3}{20}A^1 - \frac{2}{5}A^3$ $-\frac{11}{2\sqrt{21}}A^2 + G$	$-\frac{3\sqrt{15}}{20}A^1 + \frac{\sqrt{15}}{15}A^3$
		$\frac{9}{20}A^1 + \frac{2}{15}A^3$ $+\frac{1}{2\sqrt{21}}A^2 + G'$

The spin-orbit calculation for mixed configurations is a straightforward extension of the methods shown in (I). The assembly formula, which is the key part of those methods, is not modified at all for multiple shells.

We conclude our treatment of LS coupling bases by indicating some possibilities for future work.

For one thing we have always maintained a strict ordering within sub-shells, namely $1\rightarrow\ell$, $2\rightarrow\ell-1$, ... Other orderings are possible and might be useful too, particularilly for exposing inter-shell relations.

Also we have always obtained $E_{n\ n+k}$ in terms of $E_{n\ n+1}$. However, Biedenharn[32] has given a lengthy but closed algebraic expression for arbitrary k. Probably this could be made into a usable tableau formula.

Finally the parentage relations of wavefunctions resulting from one or two electrons being suddenly removed can be obtained using the assembly formula. Certain applications of this may be interesting.

C. JJ-COUPLING

The jj-coupling scheme is a useful approximation when the spin-orbit interaction, S.O., is much stronger than the electrostatic repulsion interaction, E.R.. However, the scheme also provides another approach to intermediate coupling (E.R.$\sim$S.O.) which, although mathematically equivalent to the L-S coupling approach, is operationally quite different. It will be seen that the jj-coupling scheme is characterized by many easy and straightforward operations while the L-S coupling approach involves a few difficult and complex operations. For this reason the jj-coupling approach to intermediate coupling is perhaps more suited to programing on a computer.

In jj-coupling we couple the orbital state $|{\ell \atop m}\rangle$ and the spin state $|{\frac{1}{2} \atop m_s}\rangle$ of each individual particle (Eq.57) to form states $|{j \atop m_j}\rangle$ of total angular momentum $j=\ell+\frac{1}{2},\ \ell-\frac{1}{2}$ which are eigenstates of S.O. with eigenvalues given by Eq.58.

$$|{j_1 \atop m_j}\rangle = |{\ell+\frac{1}{2} \atop m+\frac{1}{2}}\rangle = \sqrt{\frac{\ell-m}{2\ell+1}}\ |{\ell \atop m+1}\rangle\ |{\frac{1}{2} \atop -\frac{1}{2}}\rangle + \sqrt{\frac{\ell+m+1}{2\ell+1}}\ |{\ell \atop m}\rangle\ |{\frac{1}{2} \atop \frac{1}{2}}\rangle \tag{57}$$

$$|{j_2 \atop m_j}\rangle = |{\ell-\frac{1}{2} \atop m+\frac{1}{2}}\rangle = \sqrt{\frac{\ell+m+1}{2\ell+1}}\ |{\ell \atop m+1}\rangle\ |{\frac{1}{2} \atop -\frac{1}{2}}\rangle - \sqrt{\frac{\ell-m}{2\ell+1}}\ |{\ell \atop m}\rangle\ |{\frac{1}{2} \atop \frac{1}{2}}\rangle$$

$$\langle{j_1 \atop m_j}|\mathrm{S.O.}|{j_1 \atop m_j}\rangle = \frac{\ell}{2}\varepsilon = E_1 \tag{58}$$

$$\langle{j_2 \atop m_j}|\mathrm{S.O.}|{j_2 \atop m_j}\rangle = -\frac{(\ell+1)}{2}\varepsilon = E_2$$

Antisymmetric Gelfand states are now formed from the direct product of the individual particle states $|{j \atop m_j}\rangle$ to satisfy the Pauli exclusion principle. There are $2\ell+2$ different $|{}^{j_1}_{m_j}\rangle$ states and 2ℓ different $|{}^{j_2}_{m_j}\rangle$ states, so the antisymmetric Gelfand states are bases of SU($4\ell+2$). This is not surprising since we must arrive at the same bases as in the L-S coupling scheme, although by a different route. For a n-particle state comprised of n_1 $|{}^{j_1}_{m_j}\rangle$ states and n_2 $|{}^{j_2}_{m_j}\rangle$ states, the energy $\langle S.O.\rangle$ is given simply by Eq.59.

$$\langle S.O.\rangle = n_1E_1 + n_2E_2 = \mathcal{E}(n_1\ell - n_2(\ell-1))/2 \qquad (59)$$

We now wish to find eigenstates of the total angular momentum $|J\ M_j\rangle$ from these antisymmetric Gelfand states of SU($4\ell+2$). These will also be eigenstates of the electrostatic interaction operator E.R.. We encounter at this stage a type of mixed configuration $(j_1)^{n_1}(j_2)^{n_2}$ of the $|{}^{j_1}_{m_j}\rangle$ and $|{}^{j_2}_{m_j}\rangle$ states with which we are already familiar.

Indeed, we may consider the states $|{}^{j_1}_{m_j}\rangle$ to form a bases of SU($2\ell+2$) and the states $|{}^{j_2}_{m_j}\rangle$ to form a bases of SU(2ℓ) with the corresponding total angular momentum operators J_1^2 and J_2^2, so that $J^2=J_1^2+J_2^2$. For example, in the $(p)^3$ configuration we have the jj-configurations $(3/2)^3$, $(3/2)^2(1/2)$, and $(3/2)(1/2)^2$ with corresponding energies $\langle S.O.\rangle$ of $(3/2)\mathcal{E}$, $0\mathcal{E}$, and $-(3/2)\mathcal{E}$. Only the first jj-configuration is a pure configuration. The $|{}^{3/2}_{m_j}\rangle$ states form a bases of SU(4) and the $|{}^{1/2}_{m_j}\rangle$ states form a bases of SU(2). These states have the lowering operators J_1- and J_2- respectively which we may write in terms of the corresponding generators in Eq.60. We label the states $|{}^{3/2}_{3/2}\rangle$, $|{}^{3/2}_{1/2}\rangle$, $|{}^{3/2}_{-1/2}\rangle$, $|{}^{3/2}_{-3/2}\rangle$, $|{}^{1/2}_{1/2}\rangle$, and $|{}^{1/2}_{-1/2}\rangle$ as $|1\rangle, |2\rangle, \ldots, |6\rangle$.

$$\begin{aligned} J_1- &= \sqrt{3}E_{21} + 2E_{32} + \sqrt{3}E_{43} \\ J_2- &= E_{65} \end{aligned} \tag{60}$$

Altogether these states form a representation of SU(6) with lowering operator $J- = J_1- + J_2-$. We may use this lowering operator to find the antisymmetric Gelfand states of the $(3/2)^2(1/2)$ configuration in terms of the $|J\ M_J\rangle$ states by lowering the highest M_J state as shown in Eq.61.

$$\begin{aligned}
|5/2\ 5/2\rangle &= \begin{matrix}1\\2\\5\end{matrix} \\
\frac{J-}{\sqrt{5}}\,|5/2\ 5/2\rangle = |5/2\ 3/2\rangle &= \Big(\begin{matrix}1\\2\\6\end{matrix} + 2\begin{matrix}1\\3\\5\end{matrix}\Big)/\sqrt{5} \\
|3/2\ 3/2\rangle &= \Big(2\begin{matrix}1\\2\\6\end{matrix} - \begin{matrix}1\\3\\5\end{matrix}\Big)/\sqrt{5} \\
\frac{J-}{\sqrt{8}}\,|5/2\ 3/2\rangle = |5/2\ 1/2\rangle &= \Big(2\begin{matrix}1\\3\\6\end{matrix} + 3\begin{matrix}1\\4\\5\end{matrix} + 3\begin{matrix}2\\3\\5\end{matrix}\Big)/\sqrt{10} \\
\frac{J-}{\sqrt{3}}\,|3/2\ 3/2\rangle = |3/2\ 1/2\rangle &= \Big(3\begin{matrix}1\\3\\6\end{matrix} - \begin{matrix}1\\4\\5\end{matrix} - \begin{matrix}2\\3\\5\end{matrix}\Big)/\sqrt{5} \\
|1/2\ 1/2\rangle &= \Big(0 + \begin{matrix}1\\4\\5\end{matrix} - \begin{matrix}2\\3\\5\end{matrix}\Big)/\sqrt{2}
\end{aligned} \tag{61}$$

The states $|3/2\ 3/2\rangle$ and $|1/2\ 1/2\rangle$ which are orthogonal to the lowered states are found by means of Eq.11.

However, there is an alternate method for finding the angular momentum states $|J\ M_J\rangle$ in a mixed configuration which only involves the tabulation of the angular momentum states in pure configurations. We decompose the antisymmetric Gelfand states in the mixed configuration according to the scheme $SU(4\ell+2) \supset SU(2\ell+2) \times SU(2\ell) \supset SU(2) \times SU(2) \supset SU(2)$.

The decomposition of antisymmetric states of $SU(4\ell+2)$ into states of $SU(2\ell+2) \times SU(2\ell)$ is easily accomplished as shown in Eq.62 where we decompose $\begin{matrix}1\\2\\5\end{matrix}$ of SU(6) into a state of SU(4)xSU(2).

$$\begin{matrix}1\\2\\5\end{matrix} = \begin{matrix}1\\2\end{matrix} \times 5 \qquad (62)$$

The states of $SU(2\ell+2)$ and $SU(2\ell)$ will always be the antisymmetric states formed simply by splitting the column of $SU(4\ell+2)$ into two columns at the juncture of the $SU(2\ell+2)$ and $SU(2\ell)$ state labels. In the case of the more general <u>orbital</u> mixed configurations $(\ell+\ell')^n$ of Sec. 1B), we may use the decomposition $SU(2\ell+2\ell'+2) \supset SU(2\ell+1) \times SU(2\ell'+1)$ into pure configurations, or for $(\ell+\ell'+\ell'')^n$ we use the iterative decomposition $SU(2\ell+2\ell'+2\ell''+3) \supset SU(2\ell+1) \times SU(2\ell'+2\ell''+2) \supset SU(2\ell+1) \times SU(2\ell'+1) \times SU(2\ell''+1)$. However, in these cases our initial tableau need not be a single column, in which case the coefficients of such a decomposition are unknown. However, such a decomposition is always multiplicity free (no repeated rpresentations) so that such coefficients are unique and may in principle be found. In fact, much of the mathematical foundations needed to derive these coefficients are given in Sec.3.

Continuing, the decomposition $SU(2\ell+2) \supset SU(2)$ and $SU(2\ell) \supset SU(2)$ is merely a decomposition of the antisymmetric states in the pure configurations $(j_1)^{n_1}$ and $(j_2)^{n_2}$ into their respective angular momentum states $|J_1\ M_J\rangle$ and $|J_2\ M_J'\rangle$. These decompositions are given for l=1,2,3 in Table V for configurations not more than half filled. In Appendix C we show how the decomposition of the states more than half filled are related to the decompositions given in Table V. The angular momentum states given in Table V have been constructed to be eigenstates of the pairing operator P. In the $(7/2)^4$ configuration this provides the additional quantum number (seniority) necessary to distinguish between the repeated angular momentum states $|4\ M_J\rangle$ and $|2\ M_J\rangle$. The seniority number, which denotes the number of upaired particles, is given to the left of these states. Note that this is an entirely artificial classification since the pairing operator does not correspond to an approximate symmetry of the atom. However, the numbers involved in a decomposition using only Eqs.11-13 are too large to conveniently put in tabular form.

In the final step of the decomposition scheme, $SU(2)xSU(2) \supset SU(2)$, we utilize Clebsch-Gordon coefficients to reduce the direct product of angular momentum states $|J_1\ M_J\rangle \times |J_2\ M_J'\rangle$. The resultant total angular momentum states $|J\ M_J\rangle$ can then be related to the antisymmetric Gelfand states of $SU(4\ell+2)$ in a mixed configuration.

Proceeding with our example of the mixed configuration $(3/2)^2(1/2)$ of $(p)^3$, we find the total angular momentum states with M_J=J in Eq.63 from the angular momentum states $|J_1\ M_J\rangle$ and $|J_2\ M_J'\rangle$ of the $(3/2)^2$ and $(1/2)$ pure configurations.

$$|5/2\ 5/2\rangle = |22\rangle \times |\tfrac{1}{2}\tfrac{1}{2}\rangle \tag{63}$$

$$= \frac{1}{2} \times 5$$

$$= \begin{array}{c}1\\2\\5\end{array}$$

$$|3/2\ 3/2\rangle = \frac{2}{\sqrt{5}}|22\rangle \times |\tfrac{1}{2}\,-\tfrac{1}{2}\rangle - \frac{1}{\sqrt{5}}|21\rangle \times |\tfrac{1}{2}\tfrac{1}{2}\rangle$$

$$= \frac{2}{\sqrt{5}}\left(\frac{1}{2} \times 6\right) - \frac{1}{\sqrt{5}}\left(\frac{1}{3} \times 5\right)$$

$$= \left(2\begin{array}{c}1\\2\\6\end{array} - \begin{array}{c}1\\3\\5\end{array}\right)/\sqrt{5}$$

$$|1/2\ 1/2\rangle = |00\rangle \times |\tfrac{1}{2}\tfrac{1}{2}\rangle$$

$$= \frac{1}{\sqrt{2}}\left(\frac{1}{4} - \frac{2}{3}\right) \times 5$$

$$= \left(\begin{array}{c}1\\4\\5\end{array} - \begin{array}{c}2\\3\\5\end{array}\right)/\sqrt{2}$$

Similiarly, the angular momentum state $|3/2\ 3/2\rangle$ in the mixed configuration $(3/2)(1/2)^2$ is given by Eq.64 and the state $|3/2\ 3/2\rangle$ in the pure configuration $(3/2)^3$ is given by Eq.65.

$$|3/2\ 3/2\rangle = |3/2\ 3/2\rangle \times |00\rangle \tag{64}$$

$$= 1 \times \frac{5}{6}$$

$$= \begin{array}{c}1\\5\\6\end{array}$$

$$|3/2\ 3/2\rangle = \begin{array}{c}1\\2\\3\end{array} \tag{65}$$

To find the matrix elements of the electrostatic interaction, we must first transform the orbital operators V_q^k to a coupled orbit-spin bases. We first note that $V_q^k(i) \times 1_2$ acts on the uncoupled orbit-spin bases $|1\uparrow\rangle, |1\downarrow\rangle, |2\uparrow\rangle, |2\downarrow\rangle, \ldots, |2\ell+1\uparrow\rangle$, and $|2\ell+1\downarrow\rangle$ of particle i. Let U be the matrix found from Eq.57 which couples the orbit-spin bases. For example, the matrix in Eq.66 couples the orbit-spin states for p-electrons ($\ell=1$).

$$\begin{pmatrix} |{}^{3/2}_{3/2}\rangle \\ |{}^{3/2}_{1/2}\rangle \\ |{}^{3/2}_{-1/2}\rangle \\ |{}^{3/2}_{-3/2}\rangle \\ |{}^{1/2}_{1/2}\rangle \\ |{}^{1/2}_{-1/2}\rangle \end{pmatrix} = \begin{pmatrix} 1 & \cdot & \cdot & \cdot & \cdot & \cdot \\ \cdot & \sqrt{1/3} & \sqrt{2/3} & \cdot & \cdot & \cdot \\ \cdot & \cdot & \cdot & \sqrt{2/3} & \sqrt{1/3} & \cdot \\ \cdot & \cdot & \cdot & \cdot & \cdot & 1 \\ \cdot & \sqrt{2/3} & -\sqrt{1/3} & \cdot & \cdot & \cdot \\ \cdot & \cdot & \cdot & \sqrt{1/3} & -\sqrt{2/3} & \cdot \end{pmatrix} \begin{pmatrix} |1\uparrow\rangle \\ |1\downarrow\rangle \\ |2\uparrow\rangle \\ |2\downarrow\rangle \\ |3\uparrow\rangle \\ |3\downarrow\rangle \end{pmatrix} \quad (66)$$

Then W_q^k in Eq.67 becomes the operator corresponding to V_q^k when operating on products of the coupled jj bases of n particles.

$$w_q^k(i) = U(v_q^k(i) \times 1_2)U^{-1} \quad (67)$$

$$W_q^k = \sum_{i=1}^{n} w_q^k(i)$$

We may replace the V_q^k with W_q^k in Eq.33 to find the electrostatic interaction operator in our coupled orbit-spin jj bases. The electrostatic interaction energies, $\langle$E.R.$\rangle$, are then determined from matrix elements of $W^k \cdot W^k$ for even k. The operators

W^k_q are given explicitly in Table VI in terms of generators of SU(4ℓ+2) for all even k and for p,d, and f electrons. The operators are then easily evaluated when acting on the antisymmetric Gelfand bases. Using Eq.68 and Eq.34, we can then determine with the use of Table VI the matrix elements of E.R. in the jj-coupling scheme.

$$\tilde{W}^k_{-q} = (-1)^q \, W^k_q \tag{68}$$

For the $|3/2\ 3/2\rangle$ states of the $(3/2 + 1/2)^3$ jj-configurations, we have the matrix elements given by Eq.69.

$$\left\langle \begin{smallmatrix}1\\2\\3\end{smallmatrix} \right| W^2\cdot W^2 \left| \begin{smallmatrix}1\\2\\3\end{smallmatrix} \right\rangle = \left\langle \begin{smallmatrix}1\\2\\3\end{smallmatrix} \right| (W^2_0W^2_0 - W^2_{-1}W^2_1 - W^2_1W^2_{-1} + W^2_{-2}W^2_2 + W^2_2W^2_{-2}) \left| \begin{smallmatrix}1\\2\\3\end{smallmatrix} \right\rangle \tag{69}$$

$$= \left\langle \begin{smallmatrix}1\\2\\3\end{smallmatrix} \right| \tfrac{1}{6}(E_{11}-E_{22}+\sqrt{2}E_{52}-E_{33}-\sqrt{2}E_{63}+E_{44}+\sqrt{2}E_{25}-\sqrt{2}E_{36})^2 +$$

$$\tfrac{1}{6}(2E_{21}+\sqrt{3}E_{35}-\sqrt{2}E_{43}-E_{46}-E_{51}+\sqrt{3}E_{62})(\sqrt{2}E_{12}+\sqrt{3}E_{53}-\sqrt{2}E_{34}-E_{64}-E_{15}+\sqrt{3}E_{26})+$$

$$\tfrac{1}{6}(\sqrt{2}E_{12}+\sqrt{3}E_{53}-\sqrt{2}E_{34}-E_{64}-E_{15}+\sqrt{3}E_{26})(\sqrt{2}E_{21}+\sqrt{3}E_{35}-\sqrt{2}E_{43}-E_{46}-E_{51}+\sqrt{3}E_{62})+$$

$$\tfrac{1}{3}(E_{31}+E_{42}+\sqrt{2}E_{45}-\sqrt{2}E_{61})(E_{13}+E_{24}+\sqrt{2}E_{54}-\sqrt{2}E_{16})+$$

$$\tfrac{1}{3}(E_{13}+E_{24}+\sqrt{2}E_{54}-\sqrt{2}E_{16})(E_{31}+E_{42}+\sqrt{2}E_{45}-\sqrt{2}E_{61}) \left| \begin{smallmatrix}1\\2\\3\end{smallmatrix} \right\rangle$$

$$= \tfrac{5}{6} + 0 + \tfrac{3}{2} + 0 + 1 = \tfrac{10}{3}$$

$$\tfrac{1}{5} \left\langle 2\ \begin{smallmatrix}1 & 1\\2 & -3\\6 & 5\end{smallmatrix} \right| W^2\cdot W^2 \left| 2\ \begin{smallmatrix}1 & 1\\2 & -3\\6 & 5\end{smallmatrix} \right\rangle = 4/3$$

$$\left\langle \begin{smallmatrix}1\\5\\6\end{smallmatrix} \right| W^2\cdot W^2 \left| \begin{smallmatrix}1\\5\\6\end{smallmatrix} \right\rangle = 10/3$$

$$\frac{1}{\sqrt{5}}\left\langle \begin{matrix}1\\2\\3\end{matrix} \,\middle|\, W^2\cdot W^2 \,\middle|\, 2\ \begin{matrix}1&1\\2&-3\\6&5\end{matrix}\right\rangle = \sqrt{10}/3$$

$$\left\langle \begin{matrix}1\\2\\3\end{matrix} \,\middle|\, W^2\cdot W^2 \,\middle|\, \begin{matrix}1\\5\\6\end{matrix}\right\rangle = -5/3$$

$$\frac{1}{\sqrt{5}}\left\langle 2\ \begin{matrix}1&1\\2&-3\\6&5\end{matrix} \,\middle|\, W^2\cdot W^2 \,\middle|\, \begin{matrix}1\\5\\6\end{matrix}\right\rangle = \sqrt{10}/3$$

Using Eq.34 leads to the electrostatic energy submatrix Eq.70 for all $|3/2\ 3/2\rangle$ states in the jj-configurations shown.

$$\begin{matrix} & (3/2)^3 & (3/2)^2(1/2) & (3/2)(1/2)^2 \end{matrix} \qquad (70)$$

$$\langle \text{E.R.}\rangle = \begin{pmatrix} 3F^0-F^2/5 & 10F^2/25 & -F^2/5 \\ 10F^2/25 & 3F^0-11F^2/25 & 10F^2/25 \\ -F^2/5 & 10F^2/25 & 3F^0-F^2/5 \end{pmatrix}$$

Similiarly, one has the electrostatic energies (Eq.71) for the remaining states of the $(3/2)^2(1/2)$ jj-configuration.

$$\langle 1/2\ 1/2 \,|\, \text{E.R.} \,|\, 1/2\ 1/2\rangle = 3F^0 \qquad (71)$$

$$\langle 5/2\ 5/2 \,|\, \text{E.R.} \,|\, 5/2\ 5/2\rangle = 3F^0-6F^2/25$$

Adding the spin-orbit energies to the diagonal of Eq.70 gives a submatrix of the complete Hamiltonian for the electrostatic and spin-orbit interactions. The eigenstates of this submatrix are bases in the intermediate coupling scheme with eigenvalues representing the exact energy with no approximations made for the Hamiltonian considered.

2. UNITARY ANALYSIS IN MOLECULAR SYMMETRY

The tableau representations of the permutation group have been used many times before in analysis of molecular orbits. However these treatments have usually involved lengthy formulas and summations and so the tableaus are still generally regarded as being somewhat mysterious notations.[33] We shall indicate here a few of the roles which the new formulas can play in analyses that go beyond simple atomic structure. Mainly, the formulas permit one to analyse, manipulate, and evaluate operators within the n-electron states without ever involving summations over the n! permutations of electrons.

The nature of the states and operators used for representing electrons in molecules will depend a great deal upon the type of problem and degree of approximation being considered. We shall sketch two approaches to a very elementary model involving electrons orbiting three atomic cores or nuclei, viz. the structures H_3 H_3^+ H_3^{++}. We suppose at first that three orthogonal orbital states exist for a given electron. This is the same number of states that spanned the U_3 basis for the atomic p sub-shell in (I), so the mathematics will be analogous. However, the treatment of a molecular model can be quite different depending on how we choose these three states.

On one hand we could choose three molecular orbital states. These would belong to symmetry types E(x and y) and A_1 respectively, if we assume a three-fold (C_{3V}) symmetry for the nuclear arrangement. Then 2 or 3 electrons would go into these to make molecular terms as displayed on the right of Fig.12. The structure of these arrangements are quite similar to the pure and mixed configurations involving atomic orbitals.

On the other hand we could choose three more or less localized atomic orbital states around nuclei 1, 2, and 3, respectively. When 2 or 3 electrons are allowed to run in these the same terms that came from the molecular orbitals will show up as diagrammed on the right of Fig.12. However, now the states show explicitly how many electrons are on each ion, which gives us some idea of the magnitude of the coulomb repulsion energy for each state.

Before considering the multi-electron states we will try to clarify the meaning or assumed definitions that we will allow for the single electron orbital states. While doing this it should be noted that the unitary calculus is not designed to work if the single electron basis is non-orthogonal.

We consider a molecular model of some point symmetry $\mathcal{G}$ and begin by imagining that the nuclei are fixed a great distance from each other. Then the definition of separate and orthogonal atomic orbital states can be presumed. Suppose we are interested in just one orbital state $|1\rangle$ for the first nucleus and that all the other nuclei have equivalent states $|j\rangle = g_j|1\rangle$ which are obtainable by symmetry operations applied to $|1\rangle$. As long as the overlap between these states is negligible ($\langle i|j\rangle = \delta_{ij}$) there will exist an orthogonal transformation (Eq.72) between these $|j\rangle$ and molecular orbital states $|{}^{\Gamma}_{m}\rangle$ for select representations Γ of $\mathcal{G}$. (Below, $o\mathcal{G}$ is the order of group, j=1...k where ${}^{o}\mathcal{G}/k$=integer.)

$$\left|{\Gamma \atop m}\right\rangle = \frac{1}{N} P^{\Gamma}_{m.} |1\rangle = \frac{1}{N} \sum_{g} \frac{\ell^{\Gamma}}{g} D^{\Gamma\,*}_{m.}(g)\; g|1\rangle \qquad (72a)$$

$$= \sum_{j=1} |g_j\rangle\langle g_j | {\Gamma \atop m}\rangle$$

$$|g_j\rangle = \sum_{\Gamma, m} |{\Gamma \atop m}\rangle\langle{\Gamma \atop m}| g_j\rangle \qquad (72b)$$

For example, three nuclei arranged in an equilateral triangle would give the transformations shown in Eq.73. Here we used the representations of $S_3 \sim C_{3V}$ which are the same as explained in Fig.5 . The conventions followed are $(123)|1\rangle = |3\rangle$, $(123)|2\rangle = |1\rangle$ etc. for the operations and $(|{E \atop x}\rangle = |{12 \atop 3}\rangle, |{E \atop y}\rangle = |{13 \atop 2}\rangle)$ for the representations.

$$|{E \atop x}\rangle = (|1\rangle + |2\rangle - 2|3\rangle)/\sqrt{6} \qquad (73a)$$

$$|{E \atop y}\rangle = (|1\rangle - |2\rangle)/\sqrt{2}$$

$$|A_1\rangle = (|1\rangle + |2\rangle + |3\rangle)/\sqrt{3}$$

$$|1\rangle = (|{E \atop x}\rangle + \sqrt{3}\,|{E \atop y}\rangle + \sqrt{2}\,|A_1\rangle)/\sqrt{6} \qquad (73b)$$

$$|2\rangle = (|{E \atop x}\rangle - \sqrt{3}\,|{E \atop y}\rangle + \sqrt{2}\,|A_1\rangle)/\sqrt{6}$$

$$|3\rangle = (-2\,|{E \atop x}\rangle \qquad + \sqrt{2}\,|A_1\rangle)/\sqrt{6}$$

If the nuclei are brought together toward their natural separation, the unmodified atomic wavefunctions $\langle x|i\rangle$ will overlap by $S_{ij} = \langle i|j\rangle$ and be non-orthogonal and unphysical to that extent. Nevertheless, if $\mathcal{G}$ symmetry is maintained, there will always be real molecular orbital states $|{\Gamma \atop m}\rangle$ labeled by the same symmetry representations as before. Given these, we can simply <u>define</u> orthonormal localized states by the $|i\rangle$ in Eq.72b or by Eq.73b in the example.

However, it is interesting to observe that the symmetry algebra makes it possible to construct orthonormal molecular orbital states even when the $|i\rangle$ are not orthogonal. The procedure is quite convenient if no representation appears more than once. For example, the states in Eq.74 are orthonormal for all values of the overlap subject to the symmetry conditions $S_{11}=S_{22}=S_{33}$ and $S_{12}=S_{13}=S_{23}$.

$$|{E \atop x}\rangle = (|1\rangle + |2\rangle - 2|3\rangle) / \sqrt{6(S_{11}-S_{12})} \tag{74}$$

$$|{E \atop y}\rangle = (|1\rangle - |2\rangle) / \sqrt{2(S_{11}-S_{12})}$$

$$|A_1\rangle = (|1\rangle + |2\rangle + |3\rangle) / \sqrt{3(S_{11}+2S_{12})}$$

Then the states $|j\rangle$ and corresponding wavefunctions obtained by substituting Eq.74 into Eq.73b have to be orthonormal. For computational purposes such approximate definitions of $|j\rangle$ may be useful for variational calculations with explicit trial wavefunctions.

The operator calculus begins by defining the operators in terms of the states we have picked. The 1-body and 2-body operators are defined by their matrix elements $\langle i|I|j\rangle$ and $\langle ij|II|km\rangle$ in one and two particle states respectively as shown below in Eq. 75.

$$I(\alpha) \rightarrow \sum_{i,j} |i\rangle\langle i| I |j\rangle\langle j| \tag{75}$$

$$= \sum_{i,j} \langle i| I |j\rangle \, e_{ij}(\alpha)$$

$$II(\alpha,\beta) \rightarrow \sum_{i,j,k,m} \langle ij| II |km\rangle \, e_{ik}(\alpha) \, e_{jm}(\beta)$$

Then sums over the electrons give expressions that are analogous to those for atomic operators (Eq. 76).

$$\sum_{\alpha} I(\alpha) \rightarrow \sum_{i,j} \langle i| I |j\rangle \sum_{\alpha} e_{ij}(\alpha) \tag{76}$$

$$= \sum_{i,j} \langle i| I |j\rangle \, E_{ij}$$

$$\tfrac{1}{2} \sum_{\alpha \neq \beta} II(\alpha\beta) \rightarrow \tfrac{1}{2} \sum_{i,j,k,m} \langle ij| II |km\rangle \left(\sum_{\alpha,\beta} e_{ik}(\alpha) \, e_{jm}(\beta) - \sum_{\alpha} e_{ik}(\alpha) \, e_{jm}(\alpha) \right)$$

$$= \tfrac{1}{2} \sum_{i,j,k,m} \langle ij| II |km\rangle \left(E_{ik} E_{jm} - \delta_{jk} E_{im} \right)$$

The application of the operators to tableau states proceeds similarily to the atomic examples, only many more independent parameters can be expected to arise, particularily when the symmetry is low. However, various stages of approximation and semi-empirical approaches can be employed to moderate this.

For example, in the three electron problem depicted in the right side of Fig.12, there exist a quartet and two doublet states distinguished by their "covalence," and presumably, least Coulomb repulsion between the electrons. Supposing no symmetry for a moment, we imagine that the interaction between these two doublet states is done by the exchange operator II_E

(Ultimately the Coulomb operator is responsible, but we keep only the parts that connect the states.) in Eq.77 where we let $\langle ij\rangle = \langle ij|II|ji\rangle = \langle ji|II\ ij\rangle$.

$$
\begin{aligned}
II_E = {} & \tfrac{1}{2}\langle 12\rangle(E_{12}E_{21} - E_{11} + E_{21}E_{12} - E_{22}) && = \langle 12\rangle(E_{12}E_{21} - E_{11}) \\
& + \tfrac{1}{2}\langle 23\rangle(E_{23}E_{32} - E_{22} + E_{32}E_{23} - E_{33}) && + \langle 23\rangle(E_{23}E_{32} - E_{22}) \\
& + \tfrac{1}{2}\langle 13\rangle(E_{13}E_{31} - E_{11} + E_{31}E_{13} - E_{33}) && + \langle 13\rangle(E_{31}E_{13} - E_{33})
\end{aligned}
\tag{77}
$$

The matrix elements of this operator follow directly from the tableau formula (Fig.9). It is interesting to note that for these particular unitary states a double application of the Fig.9 yields the formula (Fig.7) for the permutation representation. The resulting matrix is shown in Eq.78.

$\langle II_E\rangle =$ (78)

	$\begin{smallmatrix}12\\3\end{smallmatrix}$	$\begin{smallmatrix}13\\2\end{smallmatrix}$
$\begin{smallmatrix}12\\3\end{smallmatrix}$	$\langle 12\rangle - \frac{1}{2}(\langle 23\rangle + \langle 13\rangle)$	$\frac{\sqrt{3}}{2}(\langle 23\rangle - \langle 13\rangle)$
$\begin{smallmatrix}13\\2\end{smallmatrix}$	$\frac{\sqrt{3}}{2}(\langle 23\rangle - \langle 13\rangle)$	$-\langle 12\rangle + \frac{1}{2}(\langle 23\rangle + \langle 13\rangle)$

The eigenvalues of Eq.78 shown below are the familiar results of Heitler and London.[34]

$$
\epsilon = \pm\sqrt{\langle 12\rangle^2 + \langle 23\rangle^2 + \langle 13\rangle^2 - \langle 12\rangle\langle 23\rangle - \langle 12\rangle\langle 13\rangle - \langle 23\rangle\langle 13\rangle}
\tag{79}
$$

We note that three fold symmetry ($\langle 12\rangle = \langle 23\rangle = \langle 13\rangle$) would give a degeneracy, and we prove below that the ($\begin{smallmatrix}12\\3\end{smallmatrix}$, $\begin{smallmatrix}13\\2\end{smallmatrix}$) are partners in an E doublet as shown in Fig.12. A degenerate ground state is not permitted by the Jahn Teller theorem,[35] and in fact H_3

is not even stable. Nevertheless we continue the discussion of the higher symmetry example since this gives the simplest non-trivial case of multi-electron states in point symmetry.

The derivation of states having definite point symmetry is done using the tableau assembly formula, (Fig.10) and the standard projection operators. The effect of each point group operation is seen as in Eqs.80 a and the resulting states are projected in Eqs.80b.

$$(12)\left|\begin{smallmatrix}11\\2\end{smallmatrix}\begin{smallmatrix}\uparrow\uparrow\\\downarrow\end{smallmatrix}\right\rangle = (12)\begin{smallmatrix}1\uparrow\\1\downarrow\\2\uparrow\end{smallmatrix} = \begin{smallmatrix}2\uparrow\\2\downarrow\\1\uparrow\end{smallmatrix} = \begin{smallmatrix}1\uparrow\\2\uparrow\\2\downarrow\end{smallmatrix} = -\left|\begin{smallmatrix}12\\2\end{smallmatrix}\begin{smallmatrix}\uparrow\uparrow\\\downarrow\end{smallmatrix}\right\rangle \tag{80a}$$

$$(23)\left|\begin{smallmatrix}12\\3\end{smallmatrix}\begin{smallmatrix}\uparrow\uparrow\\\downarrow\end{smallmatrix}\right\rangle = \left(\begin{smallmatrix}1\uparrow\\3\downarrow\\2\uparrow\end{smallmatrix} - \begin{smallmatrix}1\downarrow\\3\uparrow\\2\uparrow\end{smallmatrix}\right)/\sqrt{2} = \left(-\begin{smallmatrix}1\uparrow\\2\uparrow\\3\downarrow\end{smallmatrix} + \begin{smallmatrix}1\downarrow\\2\uparrow\\3\uparrow\end{smallmatrix}\right)/\sqrt{2}$$

$$= -\tfrac{1}{2}\left|\begin{smallmatrix}12\\3\end{smallmatrix}\begin{smallmatrix}\uparrow\uparrow\\\downarrow\end{smallmatrix}\right\rangle + \tfrac{\sqrt{3}}{2}\left|\begin{smallmatrix}13\\2\end{smallmatrix}\begin{smallmatrix}\uparrow\uparrow\\\downarrow\end{smallmatrix}\right\rangle$$

$$\begin{array}{lllllllll}
\sqrt{6}P^{A_1}\begin{smallmatrix}11\\2\end{smallmatrix} = & (\begin{smallmatrix}11\\2\end{smallmatrix} & -\begin{smallmatrix}12\\2\end{smallmatrix} & +\begin{smallmatrix}11\\3\end{smallmatrix} & -\begin{smallmatrix}13\\3\end{smallmatrix} & +\begin{smallmatrix}22\\3\end{smallmatrix} & -\begin{smallmatrix}23\\3\end{smallmatrix}) & /\sqrt{6} = & |A_1\rangle \qquad (80b)\\
\sqrt{6}P^{A_2}\begin{smallmatrix}11\\2\end{smallmatrix} = & (\;1 & 1 & -1 & -1 & 1 & 1) & /\sqrt{6} = & |A_2\rangle\\
\sqrt{3}P^{E}_{11}\begin{smallmatrix}11\\2\end{smallmatrix} = & (\;2 & -2 & -1 & 1 & -1 & 1) & /2\sqrt{3} = & |E_x\rangle\\
\sqrt{3}P^{E}_{21}\begin{smallmatrix}11\\2\end{smallmatrix} = & (& & 1 & -1 & -1 & 1) & /2 = & |E_y\rangle\\
\sqrt{3}P^{E}_{12}\begin{smallmatrix}11\\2\end{smallmatrix} = & (& & 1 & 1 & 1 & 1) & /2 = & |E'_x\rangle\\
\sqrt{3}P^{E}_{22}\begin{smallmatrix}11\\2\end{smallmatrix} = & (\;2 & 2 & 1 & 1 & -1 & -1) & /2\sqrt{3} = & |E'_y\rangle
\end{array}$$

$$3P^{E}_{11}\begin{smallmatrix}12\\3\end{smallmatrix} = \begin{smallmatrix}12\\3\end{smallmatrix} = |E''_x\rangle$$

$$3P^{E}_{21}\begin{smallmatrix}12\\3\end{smallmatrix} = \begin{smallmatrix}13\\2\end{smallmatrix} = |E''_y\rangle$$

In complicated computer calculations the states necessary for energy matrix calculations could be generated from a single idempotent using the trial-error-free routines described in Sec.IA. Then only one component of each degenerate term is necessary and for each of these just one row of the energy matrix needs to be calculated. The states in Eq.80b require three rows of an operator like II_D in Eq.81a to obtain the matrix below.

$$II_D = d \sum_i \sum_{j \neq i} (E_{ij}E_{ii} + E_{ii}E_{ji}) \tag{81a}$$

$$\langle II_D \rangle = d \begin{array}{c} \begin{matrix} {}^2E & {}^2E' & {}^2E'' & A_1 & A_2 \end{matrix} \\ \left[\begin{matrix} -2 & 0 & -\sqrt{\frac{3}{2}} & \cdot & \cdot \\ & 2 & \frac{3}{\sqrt{2}} & \cdot & \cdot \\ & & 0 & \cdot & \cdot \\ & & & -2 & \cdot \\ & & & & 2 \end{matrix}\right] \end{array} \tag{81b}$$

The calculations of electric or magnetic dipole transitions or electronic Raman effects involve 1-body operators and some number of parameters which can often be estimated from simple considerations of geometry and atomic structure. The treatment of spin orbit effects is the same as for the atomic case except that in general tableau states of total spin must be combined to make bases of point symmetry. Then the coupling coefficients of the point symmetry can make the desired spin-orbit states.

Clearly the tableau formulation simplifies the treatment of quantum orbital operators acting upon states of definite spin. As with the atomic orbitals it becomes possible to calculate matrix elements between states involving complicated sums over n! terms, without ever getting involved in their complexity. Indeed all manipulations of these states can be carried out using various graphical formulas.

For example, when considering the effects of part of a molecule retreating to infinity, it is convenient to ignore the exchange effects between the separated parts. Splitting a fully symmetrized tableau wavefunction into separately symmetrized parts can easily be done without involving laborious arithimetic of n! terms. One procedure employs the assembly formula (Fig. 10), as in the example below in which the symmetrization with respect to the third electron for the parent states to the right of the arrows has been dropped.

$$\left|\begin{smallmatrix}1&2\\3&\end{smallmatrix}\;\begin{smallmatrix}\uparrow&\uparrow\\,&\downarrow\end{smallmatrix}\right\rangle = \left(\begin{matrix}1\uparrow\\2\downarrow\\3\uparrow\end{matrix} - \begin{matrix}1\downarrow\\2\uparrow\\3\uparrow\end{matrix}\right)/\sqrt{2} \rightarrow \left(\begin{matrix}1\uparrow\\2\downarrow\end{matrix} - \begin{matrix}1\downarrow\\2\uparrow\end{matrix}\right)/\sqrt{2}\,\left|3\uparrow\right\rangle = \left|12\begin{smallmatrix}\uparrow\\\downarrow\end{smallmatrix}\right\rangle\left|3\uparrow\right\rangle$$

$$\left|\begin{smallmatrix}1&3\\2&\end{smallmatrix}\;\begin{smallmatrix}\uparrow&\uparrow\\,&\downarrow\end{smallmatrix}\right\rangle = -\sqrt{\frac{2}{3}}\begin{matrix}1\uparrow\\2\uparrow\\3\downarrow\end{matrix} + \sqrt{\frac{1}{6}}\begin{matrix}1\uparrow\\2\downarrow\\3\uparrow\end{matrix} + \sqrt{\frac{1}{6}}\begin{matrix}1\downarrow\\2\uparrow\\3\uparrow\end{matrix} \rightarrow -\sqrt{\frac{2}{3}}\left|\begin{smallmatrix}1\\2\end{smallmatrix}\uparrow\uparrow\right\rangle\left|3\downarrow\right\rangle + \sqrt{\frac{1}{3}}\left|\begin{smallmatrix}1\\2\end{smallmatrix}\uparrow\downarrow\right\rangle\left|3\uparrow\right\rangle$$

Generalized examples of parentage expressions like the ones above can be derived using the assembly formula or else by a more general analysis of unitary invariant operators or permutation class operators which is described in Sec. 3 below. In fact the assembly formula can be shown to follow from just this sort of analysis.

Application of these techniques to problems involving electron scattering and optical properties of reacting atoms are presently being studied.

3. FURTHER STRUCTURE OF A MULTI-PARTICLE BASIS OF THE GELFAND REPRESENTATION

In Sec. 3 of (I), we indicated that a canonical basis or Gelfand basis of U_m can be constructed from a p-rank, m-dimensional tensor space by means of the canonical projection operators of S_p as shown in Eq. 82.

$$|\mu, \nu(i_\alpha)\rangle = N_\nu \bar{P}^\lambda_{\mu\nu} \left| \begin{matrix} 1 & 2 & \cdots & p \\ i_1 & i_2 & \cdots & i_p \end{matrix} \right\rangle \tag{82}$$

The canonical structure of this Gelfand basis is determined by $\nu(i_\alpha)$ which corresponds to the unitary tableau formed by putting the state number i_α into the box of the standard tableau ν of S_p containing particle number α. We now wish to show a simple way to verify this construction using the complete set of invariant operators of U_m, and indicate applications of this theory.

It is known that the Gelfand basis of U_m is uniquely characterized by the eigenvalues of the array of Gelfand invariants given in Eq. 83.[36]

$$\begin{matrix} I^m_m & I^m_{m-1} & \cdots & I^m_1 \\ & \ddots & & \cdot^{\cdot^{\cdot}} \\ & I^2_2 & I^2_1 & \\ & & I^1_1 & \end{matrix} \tag{83}$$

The invariant operators I^ℓ_r are expressed in terms of the unit operators of U_ℓ as shown in Eq. 84.

$$I^\ell_r = \sum_{i_1 i_2 \cdots i_r}^{\ell} E_{i_1 i_2} E_{i_2 i_3} \cdots E_{i_r i_1} \tag{84}$$

$$\text{for } r = 2, 3, \ldots, \ell$$

$$I^\ell_1 = \sum_{i_1}^{\ell} E_{i_1 i_1}$$

These invariants are independent and complete when operating on the Gelfand bases.

Let p_ℓ be the number of subscript indices containing state numbers 1,2, ... ,ℓ in Eq. 82. It is clear that the eigenvalues of the I_1^ℓ are simply p_ℓ where $p_m=p$. The eigenvalues of the remaining invariant operators can be found using a closed form expression derived by Perelomov and Popov.[37]

Much effort has been made to expand the Gelfand invariants in terms of the generalized exchange operators, i.e., the class operators of S_p.[38,39] This task is accomplished by choosing the special r-cycle lower class operators $\underline{K}_r^{p_\ell}$ of $S_{(p_\ell)}$.[40] We define $S_{(p_\ell)}$ to be the subgroup of S_p corresponding to permutations of the subscript indices of our tensor bases which contain state numbers 1,2, ... ,ℓ. For example, if $|i_1 i_2 i_3 i_4 i_5\rangle = |13243\rangle$, then $S_{(p_3)}=S_{(4)}$ is the group of permutations of i_1, i_2, i_3, and i_5. Note that $S_{(4)}$ differs from S_4 where the latter is the group of permutations of the first four subscript indices i_1, i_2, i_3, and i_4. The r-cycle lower class operators $\underline{K}_r^{p_\ell}$ of $S_{(p_\ell)}$ are now defined by Eq. 85.

$$\underline{K}_r^{p_\ell} = \frac{1}{r} \overline{\sum_{j_1 j_2 \cdots j_r}} \underline{(j_1 j_2 \cdots j_r)} \qquad (85)$$

for $r=2,3,\ldots,\ell$

$$\underline{K}_1^{p_\ell} = I_1^\ell$$

The sum is restricted in Eq. 85 such that all permutations $\underline{(j_1 j_2 \cdots j_r)}$ are in the subgroup $S_{(p_\ell)}$. Also, $\underline{K}_r^{p_\ell}$ is taken to be a null operator when $p_\ell < r$. Note that our notation makes explicit the fact that the eigenvalue p_ℓ of $\underline{K}_1^{p_\ell}$ is needed to determine the subgroup $S_{(p_\ell)}$ to which the other r-cycle class operators $\underline{K}_r^{p_\ell}$ belong. It is always possible to expand the Gelfand invariants in terms of these r-cycle lower class operators, although no closed form exists. In Eq. 86 we make such an expansion for I_1^ℓ, I_2^ℓ, I_3^ℓ, and I_4^ℓ.

$$I_1^\ell = p_\ell \tag{86}$$

$$I_2^\ell = 2\underline{K}_2^{p_\ell} + \ell p_\ell$$

$$I_3^\ell = 3\underline{K}_3^{p_\ell} + 4\ell\underline{K}_2^{p_\ell} + p_\ell(p_\ell - 1) + \ell^2 p_\ell$$

$$I_4 = 4\underline{K}_4^{p_\ell} + 9\ell\underline{K}_3^{p_\ell} + (16\ell^2 + 6p_\ell - 10)\underline{K}_2^{p_\ell} + p_\ell(\ell^3 + 3\ell p_\ell - 3\ell)$$

The invariants of SU_m are simply linear combinations of the invariants of U_m[41] since the Gelfand states of U_m are also irreducible bases of SU_m. Thus the invariants of SU_m can also be expanded in terms of the r-cycle class operators. For example, the definition of the generators of SU_2 in Eq. 87 leads to the well-known Dirac identity (Eq. 88).

$$\hat{S}_z = (E_{11} - E_{22})/2 \tag{87}$$

$$\hat{S}_+ = E_{12}$$

$$\hat{S}_- = E_{21}$$

$$\hat{\mathbf{S}}^2 = I_2^2/2 - (I_1^2)^2/4 = \underline{K}_2^{p_2} + p_2 - p_2^2/4 \tag{88}$$

From Eq. 86 it follows that the r-cycle lower class operators of S_p are independent and complete invariant operators of

U_m for m=1, 2, 3, and 4. It may be proven in general that the $\underline{K}_r^{p_\ell}$ are a complete and independent set of mutually commuting Hermitian invariants of any U_m such that the eigenvalues of the array in Eq. 89 uniquely determine the Gelfand states of U_m.

$$\begin{array}{ccccccc} \underline{K}_m^{p_m} & & \underline{K}_{m-1}^{p_m} & \cdots & \underline{K}_1^{p_m} \\ & \ddots & & & \cdot^{\cdot^{\cdot}} \\ & & \underline{K}_2^{p_2} & & \underline{K}_1^{p_2} \\ & & & \underline{K}_1^{p_1} & \end{array} \tag{89}$$

Again, $\underline{K}_r^{p_\ell}$ is taken to be a null operator when $p_\ell < r$.

The fact that the r-cycle lower class operators of Eq. 89 are mutually commuting Hermitian invariant operators is very easy to demonstrate. The invariant property of the operators (Eq. 90) follows directly from the commutation of any lower permutation ($\underline{p}$) with the generators E_{ij} of U_m.

$$\left[\underline{K}_r^{p_\ell}, E_{ij}\right] = 0 \tag{90}$$

If a class contains the element (p) it also contains $(p)^{-1}$. If $(K_r^{p_\ell})^{-1}$ is the inverse of all the permutation terms in $(K_r^{p_\ell})$, then we may easily show that the r-cycle class operators are Hermitian.

$$(\underline{K}_r^{p_\ell})^\dagger = (\underline{K}_r^{p_\ell})^{-1} \tag{91}$$
$$= (\underline{K}_r^{p_\ell})$$

Finally, the class operators are mutally commuting (Eq. 92) since a class of a group commutes with all elements of that group and one of the groups $S_{(p_\ell)}$ or $S_{(p_{\ell'})}$ is a subgroup of the other.

$$\left[\underline{K}_r^{p_\ell}, \underline{K}_{r'}^{p_{\ell'}}\right] = 0 \qquad (92)$$

We may now use the r-cycle class operators to show how the canonical U_m irreducible bases are constructed from the canonical projection operators of S_p acting on the p-particle bases. We see from Eq. 93 that the projected states are eigenvectors of the operators $\underline{K}_r^{p_m}$ for all r=1,2,...,m.

$$\underline{K}_r^{p_m} \bar{P}_{\mu\nu}^{\lambda} \left| \begin{matrix} 1 & 2 & \cdots & p \\ i_1 & i_2 & \cdots & i_p \end{matrix} \right\rangle = \bar{P}_{\mu\nu}^{\lambda} \underline{K}_r^{p_m} \left| \begin{matrix} 1 & 2 & \cdots & p \\ i_1 & i_2 & \cdots & i_p \end{matrix} \right\rangle \qquad (93)$$

$$= \sum_{\nu'} D_{\nu\nu'}^{\lambda}(K_r^{p_m}) \bar{P}_{\mu\nu'}^{\lambda} \left| \begin{matrix} 1 & 2 & \cdots & p \\ i_1 & i_2 & \cdots & i_p \end{matrix} \right\rangle$$

$$= \left(\frac{N_r^{p_m} \chi_r^{\lambda}}{\ell^{\lambda}}\right) \bar{P}_{\mu\nu}^{\lambda} \left| \begin{matrix} 1 & 2 & \cdots & p \\ i_1 & i_2 & \cdots & i_p \end{matrix} \right\rangle$$

In Eq. 93, $N_r^{p_m}$ is the order of the class $\underline{K}_r^{p_m}$, χ_r^{λ} is the character of this class for the IR λ, and ℓ^{λ} is the dimension of the IR λ of S_p. This proves that the independent bases for a given μ from a bases for the IR λ of U_m (completeness of the bases will become evident later). The eigenvalues of the $\underline{K}_r^{p_m}$ given in Eq. 93 are easily evaluated in terms of a hooklength formula given in Appendix D.

We have already noted in Eq. A-8b that the projected states transform like a canonical basis ν of the IR λ under lower permutations ($\underline{p}$) of S_p. Since ν is a canonical basis it may be represented by a standard tableau of S_p. Let ν^{p_ℓ} be the IR of S_{p_ℓ} formed by removing boxes with numbers p_ℓ, $p_\ell+1$, ... , p from the standard tableau ν. Note that $\nu^{p_m} = \lambda$.

By definition, the projected states must transform like a basis of IR ν^{p_ℓ} under permutations ($\underline{p}$) of S_{p_ℓ}. In this canonical reduction of S_p, S_{p_ℓ} corresponds to the permutation of the first p_ℓ subscript indices.

Now in order to prove that the irreducible bases of U_m in Eq. 82 are canonical bases, we must show that they are eigenvectors of the remaining invariant operators $\underline{K}_r^{p_\ell}$ in Eq. 89. For $\underline{K}_r^{p_\ell}$ to be a class operator of the subgroup S_{p_ℓ} in the canonical reduction of the basis ν, it is necessary that the subgroup $S_{(p_\ell)}$ of $\underline{K}_r^{p_\ell}$ correspond to permutations of the first p_ℓ subscript indices. Since $\underline{K}_r^{p_\ell}$ permutes the state numbers 1, 2,...,ℓ, for $\underline{K}_r^{p_\ell}$ to be a class of S_{p_ℓ} in this canonical reduction these state numbers must be in the first p_ℓ subscript indices of $i_1 i_2 \ldots i_p$. This is the reason why we choose subscript indices with definite "order" such that $i_1 < i_2 < \ldots < i_p$.

Now Eq. 94 follows directly from the above considerations.

$$\underline{K}_r^{p_\ell} \, \bar{P}^{\lambda}_{\mu\nu} \left| \begin{matrix} 1 & 2 & \ldots & p \\ i_1 & i_2 & \ldots & i_p \end{matrix} \right\rangle = \frac{N_r^p \chi_r^{\nu^{p_\ell}}}{\ell^{\nu^{p_\ell}}} \, \bar{P}^{\lambda}_{\mu\nu} \left| \begin{matrix} 1 & 2 & \ldots & p \\ i_1 & i_2 & \ldots & i_p \end{matrix} \right\rangle \quad (94)$$

Thus the projected states transform like an irreducible bases of ν^{p_ℓ} of U_ℓ for $\ell = 1,2,\ldots,m$. But ν^{p_ℓ} is just the partition left after removing state numbers $\ell+1$, $\ell+2$, ... , m from the standard tableau $\nu(i_\alpha)$ of U_m. So the projected state is a canonical basis of U_m corresponding to the standard tableau $\nu(i_\alpha)$.

It is now very easy to verify the results of Appendix A. Since the invariant operators $\underline{K}_r^{p_\ell}$ are Hermitian, eigenvectors belonging to different sets of eigenvalues are orthogonal as in Eq. 95.

$$\left(\bar{P}^{\lambda}_{\mu\nu}\left|\begin{matrix}1 & 2 & \cdots & p\\ i_1 & i_2 & \cdots & i_p\end{matrix}\right\rangle,\ \bar{P}^{\lambda}_{\mu\nu'}\left|\begin{matrix}1 & 2 & \cdots & p\\ i_1 & i_2 & \cdots & i_p\end{matrix}\right\rangle\right) = 0 \qquad (95)$$

$$\text{for}\quad \nu(i_\alpha) \neq \nu'(i_\alpha)$$

Similarly, eigenvectors belonging to the same set of eigenvalues must be equal within a normalization factor as in Eq. 96.

$$\bar{P}^{\lambda}_{\mu\nu}\left|\begin{matrix}1 & 2 & \cdots & p\\ i_1 & i_2 & \cdots & i_p\end{matrix}\right\rangle = C\ \bar{P}^{\lambda}_{\mu\nu}\left|\begin{matrix}1 & 2 & \cdots & p\\ i_1 & i_2 & \cdots & i_p\end{matrix}\right\rangle \qquad (96)$$

$$\text{for}\quad \nu(i_\alpha) = \nu'(i_\alpha)$$

Not every tableau $\nu(i_\alpha)$ of U_m corresponds to a Gelfand pattern since the "betweeness conditions" (Eq. 14 of (I)) are not necessarily satisfied. However, if a tableau of U_m contains no identical state numbers in a column, it can easily be seen that the "betweeness conditions" will always be satisfied and the resulting "lexical" tableau will correspond to a Gelfand pattern. Thus the projected states with "lexical" tableaus corresponding to <u>different</u> Gelfand patterns form a complete and independent set of canonical bases of U_m. The projected states with identical state numbers in a column are orthogonal to projected states with "lexical" tableaus and must therefore be null states.

Using the r-cycle class invariants is often more convenient than using the Gelfand invariants because of the intimate connection between the permutation and unitary groups. We have already mentioned in Sec. 1.C the need for the coefficients of decomposition of $SU(2\ell+2\ell'+2)$ into the subgroup $SU(2\ell+1)$x

SU(2ℓ'+1) when dealing with mixed orbital configurations. These coefficients may easily be found with the aid of the r-cycle lower class operators.

For example, in the case of the mixed configuration $(np)^3(n'p)^3$ we know that the Gelfand states $\begin{array}{l}12\\34\\56\end{array}$ and $\begin{array}{l}12\\35\\46\end{array}$ of SU(6) are some linear combination of the states $\begin{array}{l}12\\3\end{array}\mathrm{x}\begin{array}{l}45\\6\end{array}$ and $\begin{array}{l}12\\3\end{array}\mathrm{x}\begin{array}{l}46\\5\end{array}$ of SU(3)xSU(3). To find this combination we need only to simultaneously diagonalize the r-cycle class operators in Eq. 97 in the Gelfand bases of SU(6).

$$1 \times \begin{pmatrix} \underline{K}_3^3 & & \underline{K}_2^3 & & \underline{K}_1^3 \\ & \underline{K}_2^2 & & \underline{K}_1^2 & \\ & & \underline{K}_1^1 & & \end{pmatrix} \qquad (97)$$

The classes in Eq. 97 only permute the indices i_4, i_5, and i_6. Fortunately, for states with only two columns it is always sufficient to use only the unicycle and bicycle class operators. The Gelfand states of SU(6) are already eigenvectors of the unicycle operators of Eq. 97 which are given in Eq. 98.

$$\begin{aligned} 1\mathrm{x}\underline{K}_1^3 &= E_{44}+E_{55}+E_{66} \\ 1\mathrm{x}\underline{K}_1^2 &= E_{44}+E_{55} \\ 1\mathrm{x}\underline{K}_1^1 &= E_{44} \end{aligned} \qquad (98)$$

Of the two operators remaining, it is sufficient to use the operator $1\mathrm{x}\underline{K}_2^2 = (\underline{45})$ to determine the coefficients of decomposition in our case. Eq. 99 shows the operator $(\underline{45})$ in the Gelfand bases of SU(6).

$$\begin{array}{ccc} & \begin{matrix}12\\34\\56\end{matrix} & \begin{matrix}12\\35\\46\end{matrix} \\ (\underline{45}) = & \begin{pmatrix} -1/2 & \sqrt{3}/2 \\ \sqrt{3}/2 & 1/2 \end{pmatrix} & \end{array} \qquad (99)$$

Diagonalizing the operator ($\underline{45}$) leads to the decompositions in Eq. 100 where $\begin{smallmatrix}12\\3\end{smallmatrix} \times \begin{smallmatrix}45\\6\end{smallmatrix}$ and $\begin{smallmatrix}12\\3\end{smallmatrix} \times \begin{smallmatrix}46\\5\end{smallmatrix}$ have eigenvalues 1 and -1 respectively under ($\underline{45}$).

$$\begin{matrix}12\\34\\56\end{matrix} = 1/2 \; \begin{smallmatrix}12\\3\end{smallmatrix} \times \begin{smallmatrix}45\\6\end{smallmatrix} + \sqrt{3}/2 \; \begin{smallmatrix}12\\3\end{smallmatrix} \times \begin{smallmatrix}46\\5\end{smallmatrix} \qquad (100)$$

$$\begin{matrix}12\\35\\46\end{matrix} = \sqrt{3}/2 \; \begin{smallmatrix}12\\3\end{smallmatrix} \times \begin{smallmatrix}45\\6\end{smallmatrix} - 1/2 \; \begin{smallmatrix}12\\3\end{smallmatrix} \times \begin{smallmatrix}46\\5\end{smallmatrix}$$

From this example it can again be seen that the explicit construction of Gelfand states using the canonical projection operators of the symmetric group has more than academic value. It provides another means of analysing the unitary group using the older and better developed theory of the symmetric group.

APPENDIX A: Permutations and Independent Gelfand Bases

Two commuting groups $(\bar{p}\ldots)$ and $(\underline{p}\ldots)$ of permutations were defined in Sec. 3. The first type $\bar{p}$ moves specific electrons form state to state as for example in Eq.A-1.

$$(\overline{124}) \left|\begin{matrix}1234\\ \mathrm{abcd}\end{matrix}\right\rangle = \left|\begin{matrix}4132\\ \mathrm{abcd}\end{matrix}\right\rangle = \left|\begin{matrix}1234\\ \mathrm{bdca}\end{matrix}\right\rangle \qquad \text{(A-1)}$$

The second type $\underline{p}$ moves specific states form electron to electron as in Eq.A-2.

$$(\underline{124}) \left|\begin{matrix}\alpha\,\beta\,\gamma\,\delta\\ i_1 i_2 i_3 i_4\end{matrix}\right\rangle = \left|\begin{matrix}\alpha\,\beta\,\gamma\,\delta\\ i_4 i_1 i_3 i_2\end{matrix}\right\rangle \qquad \text{(A-2)}$$

Since only the <u>relative</u> position of electrons and states has any meaning, there is a relation between each electron permutation $\bar{p}$ and the corresponding state permutation $\underline{p}$. The action of $\underline{p}$ on the state $\left|\begin{matrix}1\;2\\ i_1 i_2\ldots\end{matrix}\right\rangle$ of unpermuted order is equal to the action of the corresponding inverse: $(\bar{p})^{-1}$.

$$\underline{p} \left|\begin{matrix}1\;2\;\ldots n\\ i_1 i_2 \ldots i_n\end{matrix}\right\rangle = (\bar{p})^{-1} \left|\begin{matrix}1\;2\;\ldots n\\ i_1 i_2 \ldots i_n\end{matrix}\right\rangle \qquad \text{(A-3)}$$

To find the action of $\underline{p}$ on a general state $\bar{m}\left|\begin{matrix}1\;2\;\ldots n\\ i_1 i_2 \ldots i_n\end{matrix}\right\rangle$, we use the fact that $\underline{p}$ must commute with every $\bar{m}$.

$$\begin{aligned}\underline{p}\,\bar{m}\left|\begin{matrix}1\;2\;\ldots n\\ i_1 i_2 \ldots i_n\end{matrix}\right\rangle &= \bar{m}\,\underline{p}\left|\begin{matrix}1\;2\;\ldots n\\ i_1 i_2 \ldots i_n\end{matrix}\right\rangle \\ &= \bar{m}\,\bar{p}^{-1}\left|\begin{matrix}1\;2\;\ldots n\\ i_1 i_2 \ldots i_n\end{matrix}\right\rangle = \bar{m}\,\bar{p}^{-1}\bar{m}^{-1}\bar{m}\left|\begin{matrix}1\;2\;\ldots n\\ i_1 i_2 \ldots i_n\end{matrix}\right\rangle\end{aligned} \qquad \text{(A-4)}$$

Now the group algebra of S_n is expressed by the properties of the projectors (Eq.A-5a), namely completeness (Eq.A-5b) and orthonormality (Eq.A-5c).

$$P^{\lambda}_{\mu\nu} = \frac{\ell^{\lambda}}{n!} \sum_{p} D^{\lambda}_{\mu\nu}(p)\, p \qquad \text{(A-5a)}$$

$$p = \sum_{\lambda'} \sum_{\mu'\nu'} D^{\lambda'}_{\mu'\nu'}(p)\, P^{\lambda'}_{\mu'\nu'} \qquad \text{(A-5b)}$$

$$P^{\lambda}_{\mu\nu}\; P^{\lambda'}_{\mu'\nu'} = \delta^{\lambda'\lambda}\, \delta_{\nu\mu'}\, P^{\lambda}_{\mu\nu'} \qquad \text{(A-5c)}$$

The Yamanouchi real orthogonal representations of Fig.7 are used.This algebra and the resulting transformation rules (Eqs.A-6) apply either to the group of ($\bar{p}$...) or of ($\underline{p}$...) since these two are mathematically identical.

$$p\, P^{\lambda}_{\mu\nu} = \sum_{\mu'} D^{\lambda}_{\mu'\mu}(p)\, P^{\lambda}_{\mu'\nu} \qquad \text{(A-6a)}$$

$$P^{\lambda}_{\mu\nu}\; p = \sum_{\nu'} D^{\lambda}_{\nu\nu'}(p)\, P^{\lambda}_{\mu\nu'} \qquad \text{(A-6b)}$$

However, we shall use only the projectors made from ($\bar{p}$...) because the scalar product is defined with respect to electron numbers and only the $\bar{p}$ obeys the unitary conditions of Eqs.A-7 in general, while the $\underline{p}$ may not be unitary.

$$(|x\rangle,\ \bar{p}\,|y\rangle) = \langle x|\bar{p}|y\rangle = (\bar{p}^{-1}\,|x\rangle,\ |y\rangle) \qquad \text{(A-7a)}$$

$$(|x\rangle,\ \bar{P}^{\lambda}_{\mu\nu}\,|y\rangle) = \langle x|\,\bar{P}^{\lambda}_{\mu\nu}\,|y\rangle = (\bar{P}^{\lambda}_{\nu\mu}\,|x\rangle,\ |y\rangle) \qquad \text{(A-7b)}$$

A-5a + A-5b $\Rightarrow$ $\frac{\ell^{\lambda}}{n!} \sum_{p} D^{\lambda}_{\mu\nu}(p)\, D^{\lambda'}_{\alpha\beta}(p) = \delta_{\mu\alpha}\,\delta_{\nu\beta}\,\delta_{\lambda\lambda'}$

Now the effect of the two kinds of permutations on the multi-particle states can be seen.

$$\bar{p}|\mu,\nu\rangle = N_\nu \; \bar{p}\, \bar{P}^{\lambda}_{\mu\nu} \left| \begin{matrix} 1 & 2 & \ldots n \\ i_1 & i_2 & \ldots i_n \end{matrix} \right\rangle = \sum_{\mu'} D^{\lambda}_{\mu'\mu}(p) \quad |\mu',\nu\rangle \tag{A-8a}$$

$$\begin{aligned} \underline{p}\,|\mu,\nu\rangle &= N_\nu \; \bar{P}^{\lambda}_{\mu\nu} \; \underline{p} \left| \begin{matrix} 1 & 2 & \ldots n \\ i_1 & i_2 & \ldots i_n \end{matrix} \right\rangle \\ &= N_\nu \; \bar{P}^{\lambda}_{\mu\nu} \; \bar{p}^{-1} \left| \begin{matrix} 1 & 2 & \ldots n \\ i_1 & i_2 & \ldots i_n \end{matrix} \right\rangle \\ &= \sum_{\nu'} D^{\lambda}_{\nu'\nu}(p) \; |\mu,\nu'\rangle \end{aligned} \tag{A-8b}$$

This last equation can be used to deduce all possible restrictions on the ν tableaus. For example, we observe that the projection in Eq.A-9, which would try to put two equal unitary states in the same column of $\nu(i_\alpha)$, is zero.

$$\begin{aligned} P_{\mu\, \begin{smallmatrix}134\\2\end{smallmatrix}} \left| \begin{matrix} 1234 \\ 5556 \end{matrix} \right\rangle &= (\underline{12})\, \bar{P}_{\mu\, \begin{smallmatrix}134\\2\end{smallmatrix}} \left| \begin{matrix} 1234 \\ 5556 \end{matrix} \right\rangle \\ &= D_{\begin{smallmatrix}134\\2\end{smallmatrix}\, \begin{smallmatrix}134\\2\end{smallmatrix}}(12) \quad \bar{P}_{\mu\, \begin{smallmatrix}134\\2\end{smallmatrix}} \left| \begin{matrix} 1234 \\ 5556 \end{matrix} \right\rangle \\ &= -\bar{P}_{\mu\, \begin{smallmatrix}134\\2\end{smallmatrix}} \left| \begin{matrix} 1234 \\ 5556 \end{matrix} \right\rangle \end{aligned} \tag{A-9}$$

Similarily, another projection shown in Eq.A-10 is seen to be a non-zero multiple of the one in Eq.A-9 and is likewise to be thrown out.

$$\bar{P}_{\mu\,\substack{124\\3}} \left|\begin{matrix}1234\\5556\end{matrix}\right\rangle = (\underline{23})\, \bar{P}_{\mu\,\substack{124\\3}} \left|\begin{matrix}1234\\5556\end{matrix}\right\rangle \qquad \text{(A-10)}$$

$$= \left(D(23)_{\substack{124\\3}\;\substack{124\\3}} \; \bar{P}_{\mu\,\substack{124\\3}} + D(23)_{\substack{134\\2}\;\substack{124\\3}} \; \bar{P}_{\mu\,\substack{134\\2}} \right) \left|\begin{matrix}1234\\5556\end{matrix}\right\rangle$$

$$= \left(-1/2\; \bar{P}_{\mu\,\substack{124\\3}} + \sqrt{3}/2\; \bar{P}_{\mu\,\substack{134\\2}} \right) \left|\begin{matrix}1234\\5556\end{matrix}\right\rangle$$

$$\bar{P}_{\mu\,\substack{124\\3}} \left|\begin{matrix}1234\\5556\end{matrix}\right\rangle = 1/\sqrt{3}\; \bar{P}_{\mu\,\substack{134\\2}} \left|\begin{matrix}1234\\5556\end{matrix}\right\rangle$$

Indeed, this procedure is quite general. All identity permutations p^I ($\underline{p}^I | i_1 i_2 \ldots i_n \rangle = | i_1 i_2 \ldots i_n \rangle$) are products of transpositions (m, $m \pm 1$) because of the initial state ordering $i_1 \leqslant i_2 \leqslant \ldots \leqslant i_n$. Now if a ν differs from a ν by an ($m, m \pm 1$), the corresponding projections differ only by a non-zero proportionality constant $\sqrt{d - 1} / \sqrt{d + 1}$ as in A-10. This constant is derived from Fig. 7. Indeed it can be seen that each of the ν' arising from ν by a p^I need not be considered further after the ν projection is made. When $d = 1$ the operation of (m, $m \pm 1$) gives the projection back again with a factor of (+1) or (-1) if the ($m - 1$) box is on the left side or on top of the m box in ν, respectively. For the latter case the ν projection, and all those proportional to it, are thrown out as was the case in Eqs. A-9 and A-10.

The normalization of these states is defined by Eqs.A-11, where Eqs.A-7b, A-5a, and A-5c have been used.

$$\left(N_\nu \, P^\lambda_{\mu\nu} \left| \begin{smallmatrix} 1 & 2 & \dots & n \\ i_1 & i_2 & \dots & i_n \end{smallmatrix} \right\rangle , \; N_\nu \, P^\lambda_{\mu\nu} \left| \begin{smallmatrix} 1 & 2 & \dots & n \\ i_1 & i_2 & \dots & i_n \end{smallmatrix} \right\rangle \right) = 1 \qquad \text{(A-11)}$$

$$N_\nu^{\,2} \left\langle \begin{smallmatrix} 1 & 2 & \dots & n \\ i_1 & i_2 & \dots & i_n \end{smallmatrix} \right| P^\lambda_{\nu\mu} \, P^\lambda_{\mu\nu} \left| \begin{smallmatrix} 1 & 2 & \dots & n \\ i_1 & i_2 & \dots & i_n \end{smallmatrix} \right\rangle = 1$$

$$N_\nu^{\,2} \, \frac{\ell^\lambda}{n!} \sum_p D^\lambda_{\nu\nu}(p) \left\langle \begin{smallmatrix} 1 & 2 & \dots & n \\ i_1 & i_2 & \dots & i_n \end{smallmatrix} \right| \bar{p} \left| \begin{smallmatrix} 1 & 2 & \dots & n \\ i_1 & i_2 & \dots & i_n \end{smallmatrix} \right\rangle = 1$$

Because of the individual state normilization, only the identity permutations p^I survive in the above. Eq.44 is then proved.

APPENDIX B: Permutations and Unitary Operators

Here we give an example in which the Yamanouchi S_n representation tableau formula (Fig.7) is used to find a U_m representation element given by our tableau formula in Fig.9. This should help to explain the relation between U_m and S_n. A complete proof of the Gelfand - S_n correspondence involves relations[42] between certain classes of permutations $\underline{p}$ and the U_m Casimir invariants used by Biedenharn and Louck[43] to define the Gelfand representation. These class relations will be discussed in Part II where their computational applications are shown.

Consider the operator E_{23} acting on the three particle state shown in Eq.B-1, where Eqs.11 and 39 are used for the operator and state, respectively.

$$E_{23}\left|\mu,\ \begin{smallmatrix}13\\3\end{smallmatrix}\right\rangle = \left(\sum_{\alpha=1}^{3} e_{23}(\alpha)\right)\left(N_1\ \bar{P}^{210}_{\mu\begin{smallmatrix}12\\3\end{smallmatrix}} \left|\begin{smallmatrix}123\\133\end{smallmatrix}\right\rangle\right) \tag{B-1}$$

Now $e_{23}(\alpha)$ is defined to commute with all permutation operators and to replace a 3-state with a 2-state for any electron α that is in a 3-state.

$$E_{23}\left|\mu,\ \begin{smallmatrix}13\\3\end{smallmatrix}\right\rangle = N_1\ \bar{P}^{210}_{\mu\begin{smallmatrix}12\\3\end{smallmatrix}} \left(\left|\begin{smallmatrix}123\\123\end{smallmatrix}\right\rangle + \left|\begin{smallmatrix}123\\132\end{smallmatrix}\right\rangle\right) \tag{B-2}$$

$$= N_1\ \bar{P}^{210}_{\mu\begin{smallmatrix}12\\3\end{smallmatrix}} \left((\underline{1}) + (\underline{23})\right)\left(\left|\begin{smallmatrix}123\\123\end{smallmatrix}\right\rangle\right)$$

The effect of this operation is expressed in terms of permutations in Eq.B-2 and expanded using the Yamanouchi S_n formula below.

$$E_{23}\left|\mu,\ \begin{smallmatrix}13\\3\end{smallmatrix}\right\rangle = N_1\left[\left(D_{\begin{smallmatrix}12\\3\end{smallmatrix}\begin{smallmatrix}12\\3\end{smallmatrix}}(1) + D_{\begin{smallmatrix}12\\3\end{smallmatrix}\begin{smallmatrix}12\\3\end{smallmatrix}}(23)\right)\bar{P}_{\mu\begin{smallmatrix}12\\3\end{smallmatrix}} + \left(D_{\begin{smallmatrix}12\\3\end{smallmatrix}\begin{smallmatrix}13\\2\end{smallmatrix}}(1) + D_{\begin{smallmatrix}12\\3\end{smallmatrix}\begin{smallmatrix}13\\2\end{smallmatrix}}(23)\right)\bar{P}_{\mu\begin{smallmatrix}13\\2\end{smallmatrix}}\right]\left|\begin{smallmatrix}123\\123\end{smallmatrix}\right\rangle \tag{B-3}$$

$$= \left[\frac{N_1}{N_1'}\left(1-\tfrac{1}{2}\right)N_1'\ \bar{P}_{\mu\begin{smallmatrix}12\\3\end{smallmatrix}} + \frac{N_1}{N_1''}\left(0+\sqrt{\tfrac{3}{2}}\right)N_1''\ \bar{P}_{\mu\begin{smallmatrix}13\\2\end{smallmatrix}}\right]\left|\begin{smallmatrix}123\\123\end{smallmatrix}\right\rangle$$

Evaluation of the normalizations using Eq.44 gives the results (Eq.B-4) which corresponds to the example in Fig.9-e.

$$E_{23}\left|\mu,\ \begin{smallmatrix}13\\3\end{smallmatrix}\right\rangle = \sqrt{\tfrac{1}{2}}\left|\mu,\ \begin{smallmatrix}12\\3\end{smallmatrix}\right\rangle + \sqrt{\tfrac{3}{2}}\left|\mu,\ \begin{smallmatrix}13\\2\end{smallmatrix}\right\rangle \tag{B-4}$$

APPENDIX C: More Than Half Filled Subshells

A more than half filled subshell is a pure configuration $(\ell)^{n'}$ for which $n' > 2\ell+1$. The total angular momentum terms for such a more than half filled (MHF) subshell are the same as for the less than half filled (LHF) subshell $(\ell)^n$ for which $n=4\ell+2-n'$. This result can be understood most clearly in terms of the associated Gelfand bases of $SU(2\ell+1)$ which we now define.

Let the numbered Young tableau $|a\rangle$ represent a Gelfand basis of $SU(2\ell+1)$. Since we will be dealing with orbital states, we limit $|a\rangle$ to tableaus of one and two columns. Since no column may have a number repeated twice, $|a\rangle$ must contain each of the numbers $1,2,\ldots,2\ell+1$ twice or less. We may now define $|a^*\rangle$, the Gelfand basis associated with $|a\rangle$, by the following prescriptions:

a) Any number contained once in $|a\rangle$ is contained once in $|a^*\rangle$ in the same column.

b) Any number not contained in $|a\rangle$ is contained twice in $|a^*\rangle$ in different columns.

c) Any number contained twice in $|a\rangle$ is not contained in $|a^*\rangle$.

d) The numbers are ordered lexically so that they increase downward in each column.

The prescription d) is automatically fulfilled by following prescriptions a) to c) with the successive numbers $1,2,\ldots,2\ell+1$. This shows that $|a^*\rangle$ is unique. Also it is easily seen that $|a^{**}\rangle = |a\rangle$.

In Eq.C-1 we give an example for $\ell=2$.

$$\begin{matrix} 1 & 2 \\ 2 & \\ 4 & \end{matrix}^{*} = \begin{matrix} 1 & 3 \\ 3 & 5 \\ 4 & \\ 5 & \end{matrix} \qquad (C-1)$$

We show in Fig.13 an equivalent means of finding the associated basis given by Baird and Biedenharn.[44] The method shown is more general since $|a\rangle$ need not be restricted to two columns.

It is now possible to relate the matrix elements of the electronic orbital operators $E_{i-1\ i}$ in the Gelfand basis to those in the associated Gelfand basis via Eq.C-2.

$$\langle a | E_{i-1\ i} | b \rangle = \langle b^{*} | E_{i-1\ i} | a^{*} \rangle \qquad (C-2)$$

This relation is valid for all Gelfand bases.[45] However, if we restrict our attention to Gelfand states of two columns, the proof of Eq.C-2 is a simple result of applying the prescriptions a) to d) to the matrix elements shown in Fig.10. The application of these prescriptions to Eq. c) in Fig.10 is shown in Fig.14 to verify Eq.C-2. Similar applications of these prescriptions to the remaining equations in Fig.10 verifies Eq.C-2 for all possible matrix elements, and thus constitutes a proof in our limited case.

Using the commutations relations for the orbital operators E_{ij} and Eq.C-2 results in Eq.C-3, which relates the matrix elements of any orbital operator E_{ij} in the Gelfand basis to those in the associated Gelfand basis.

$$\langle a | E_{ij} | b \rangle = (-1)^{i-j+1} \langle b^* | E_{ij} | a^* \rangle \qquad (C\text{-}3)$$

$$= (-1)^{i-j+1} \langle a^* | E_{ji} | b^* \rangle$$

The operator L_z is a linear combination of the diagonal orbital operators E_{ii} with eigenvalues M_L. From Eq.C-3 it follows that the total orbital angular momentum states $|L\ M_L\rangle$ of $(\ell)^n$ are a linear combination of Gelfand states $|a\rangle$ such that the associated orbital state $|(L\ M_L)^*\rangle$ has an eigenvalue $-M_L$ of L_z. There is a one to one correspondence between the Gelfand bases $|a\rangle$ of $(\ell)^n$ and their associated bases of $(\ell)^{n'}$. Also, the spectra of eigenvalues M_L in a configuration completely determine the total angular momentum terms it contains . Thus the total angular momentum terms conained in the MHF subshell $(\ell)^{n'}$ are the same as in the LHF subshell $(\ell)^n$ for which $n=4\ell+2-n'$. This fact leads us to investigate the relationship of single and double tensor operator matrix elements in the $(\ell)^{n'}$ configuration to those in the $(\ell)^n$ configuration using the idea of the associated Gelfand bases.

Single Tensor Operators in MHF Subshells

We first note that the single tensor orbital operators V^k_q are a linear combination of the orbital generators E_{ij} with $q+j=i$ as shown in Eq.C-4 and Eq.C-5.

$$V^k_q = (\ell i | V^k_q | \ell j)\ E_{ij} \qquad (C\text{-}4)$$

$$V^k_{-q} = (-1)^q (\ell i | V^k_q | \ell j)\ E_{ji} \qquad (C\text{-}5)$$

From Eqs. C-3, C-4, and C-5 we derive Eq. C-6 which relates single tensor operators in the $(\ell)^n$ configuration to those in the $(\ell)^{n'}$ configuration.

$$\langle a \mid V_q^k \mid b \rangle \quad = \quad - \langle a^* \mid V_{-q}^k \mid b^* \rangle \tag{C-6}$$

We may expand the total orbital angular momentum operator $\hat{L}^2$ in terms of the V_q^k's as shown in Eq.C-7.

$$\hat{L}^2 \quad = \quad \alpha \sum_{q=-1}^{1} (-1)^q \, V_q^1 \, V_{-q}^1 \tag{C-7}$$

$$\alpha = \frac{\ell(\ell+1)(2\ell+1)}{3}$$

Using Eqs.C-6 and C-7 we may now relate the total orbital angular momentum bases $|(\ell^n \, L \, M_L)\rangle$ in the LHF subshell to the bases $|(\ell^{n'} \, L \, M_L)\rangle$ in the MHF subshell by means of the associated bases shown in Eq.C-8.

$$|(\ell^n \, L \, M_L)^*\rangle \quad = \quad |(\ell^{n'} \, L \, -M_L)\rangle \tag{C-8}$$

For example, in Eq.C-9 we show the relationship between the bases of the $(p)^2$ and $(p)^4$ configuration for which L=1.

$$|P\ 1\rangle = \left|\begin{smallmatrix}1\\2\end{smallmatrix}\right\rangle \qquad |P\ 0\rangle = \left|\begin{smallmatrix}1\\3\end{smallmatrix}\right\rangle \qquad |P\ -1\rangle = \left|\begin{smallmatrix}2\\3\end{smallmatrix}\right\rangle \tag{C-9}$$

$$|P\ -1\rangle = \left|\begin{smallmatrix}13\\2\\3\end{smallmatrix}\right\rangle \qquad |P\ 0\rangle = \left|\begin{smallmatrix}12\\2\\3\end{smallmatrix}\right\rangle \qquad |P\ 1\rangle = \left|\begin{smallmatrix}11\\2\\3\end{smallmatrix}\right\rangle$$

For self-associated bases where n=n' we have a convenient means of finding the $|L \, -M_L\rangle$ bases from the $|L \, M_L\rangle$ bases. Thus, in Eq.C-10 we may find $|P \, -1\rangle$ of the $(p)^3$ configuration by taking the associated bases of $|P\ 1\rangle$.

$$|P\ 1\rangle = \left(\left|{}^{12}_{2}\right\rangle - \left|{}^{11}_{3}\right\rangle \right)/\sqrt{2}$$

$$|P\ {-1}\rangle = \left(\left|{}^{13}_{3}\right\rangle - \left|{}^{22}_{3}\right\rangle \right)/\sqrt{2} \qquad \text{(C-10)}$$

We now derive the relation (Eq.C-13) between the reduced matrix elements of the orbital operators V^k_q in MHF and LHF sub-shells. First we find Eq.C-11 directly from Eqs.C-6 and C-8.

$$\langle \ell^n\ L\ M_L \,|\, V^k_q \,|\, \ell^n\ L'\ M_L' \rangle = -\langle \ell^{n'}\ L\ {-M_L} \,|\, V^k_{-q} \,|\, \ell^{n'}\ L'\ {-M_L'} \rangle \qquad \text{(C-11)}$$

Using the definition of the reduced matrix elements (Eq.C-12) and Eq.C-11, we may easily derive Eq.C-13.

$$\langle \ell^n\ L\ M_L \,|\, V^k_q \,|\, \ell^n\ L'\ M_L' \rangle = (-1)^{L-M_L} \sqrt{(2k+1)} \begin{pmatrix} L & k & L' \\ -M_L & q & M_L' \end{pmatrix} \times (\ell^n\ L \,\|\, V^k \,\|\, \ell^n\ L') \qquad \text{(C-12)}$$

$$(\ell^n\ L \,\|\, V^k \,\|\, \ell^n\ L') = (-1)^{k+1+L+L'} (\ell^{n'} L \,\|\, V^k \,\|\, \ell^{n'} L') \qquad \text{(C-13)}$$

Eq.C-13 would agree with Racah's result,[46] where the factor $(-1)^{L+L'}$ is unity, if we let Eq.C-14 define the associated bases instead of Eq.C-8.

$$|(\ell^n\ L\ M_L)^*\rangle = (-1)^L\ |(\ell^{n'}\ L\ {-M_L})\rangle \qquad \text{(C-14)}$$

This is possible since the choice of sign is arbitrary; however, we shall continue to use the simpler Eq.C-8.

Double Tensor Operators in MHF Subshells

We wish to find matrix elements of double tensor operators in a basis where orbital states of SU(2ℓ+1) and spin states of SU(2) are coupled to form antisymmetric states of SU(4ℓ+2). These coupling coefficients have been determined in Fig.10 We may use these coupling coefficients in conjunction with prescriptions a) to d) to relate the coupling coefficients of the orbital and spin states to those of their associated bases. First it is necessary to define the associated bases for SU(2) spin states and antisymmetric SU(4ℓ+2) states.

We define the associated bases of the SU(4ℓ+2) states in accordance with the general method shown in Fig.13. The associated basis of the antisymmetric SU(4ℓ+2) state uniquely determines M_S of the associated spin state. Since the associated spin state is conjugate to the associated orbital basis, S is also uniquely defined. In Fig.15 we show the construction of associated bases of antisymmetric SU(4ℓ+2) states and SU(2) spin states. We note that $|(S\ M_S)^*\rangle = |S\ -M_S\rangle$ as in the orbital case.

From prescription a) we find, using obvious notation, that the associated coupling coefficient factors are related as shown in Eq.C-15a. Similarly, Eq.C-15b - C-15c follow from prescriptions b) and c) respectively, where we define $(u_1 u_2|0)=1$ as the coupling coefficient factor when the number $n=u_1+u_2$ is not contained in the SU(2ℓ+1) orbital state.

$$(u_1|\uparrow)^* = (u_1|\downarrow) \qquad (u_1|\downarrow)^* = (u_1|\uparrow)$$
$$(u_2|\uparrow)^* = -(u_2|\downarrow) \qquad (u_2|\downarrow)^* = -(u_2|\uparrow) \tag{C-15a}$$

$$(u_1u_2|\uparrow\downarrow)^* = (-1)^{u_1+u_2}\,(u_1u_2|0) \tag{C-15b}$$

$$(u_1u_2|0)^* = (-1)^{u_1+u_2}\,(u_1u_2|\uparrow\downarrow) \tag{C-15c}$$

It follows that the associated coupling coefficients are the same as the coupling coefficients except for a sign change $(-1)^N$ given by Eq.C-16 which is determined solely from the orbital or SU($2\ell+1$) basis $|a\rangle$.

$$(-1)^N = \prod_{a,b,c} (-1)^{N_a+N_b+N_c} \tag{C-16}$$

The product in Eq.C-16 is over all numbers a contained once in $|a\rangle$, all numbers b not contained in $|a\rangle$, and all numbers c contained twice in $|a\rangle$. We define N_a, N_b, and N_c in Eq.C-16 as follows: $N_a=0$ if a is in column 1 and $N_a=1$ if a is in column 2; N_b is the number of boxes containing numbers less than b; and N_c is the number of boxes containing numbers less than c.

An example is shown in Eq.C-17 for the associated orbital basis given previously in Eq.C-1.

$$\left\langle \begin{array}{l} 12 \\ 2 \\ 4 \end{array} \begin{array}{l} \uparrow\uparrow\downarrow \\ \downarrow \\ \end{array} \middle| \begin{array}{l} 1\uparrow \\ 2\uparrow \\ 2\downarrow \\ 4\downarrow \end{array} \right\rangle = (-1)^{0+7+1} \left\langle \begin{array}{l} 12^* \\ 2 \\ 4 \end{array} \begin{array}{l} \uparrow\uparrow\downarrow^* \\ \downarrow \\ \end{array} \middle| \begin{array}{l} 1\uparrow \\ 2\uparrow \\ 2\downarrow \\ 4\downarrow \end{array} \right\rangle \tag{C-17}$$

$$= \left\langle \begin{array}{l} 13 \\ 35 \\ 4 \\ 5 \end{array} \begin{array}{l} \uparrow\uparrow\uparrow\downarrow \\ \downarrow\downarrow \\ \\ \end{array} \middle| \begin{array}{l} 1\downarrow \\ 3\uparrow \\ 3\downarrow \\ 4\uparrow \\ 5\uparrow \\ 5\downarrow \end{array} \right\rangle$$

In this example, $N_1=0$ and $N_4=0$; $N_3=3$ and $N_5=4$; and $N_2=1$.

A direct consequence of Eq.C-16 is that the sign change $(-1)^N$ of the coupling coefficient of the associated bases alternates with M_L of the orbital state for a given S. That is, for a given S, $(-1)^N$ changes sign whenever one box in a Gelfand state of SU($2\ell+1$) is raised from n to n-1 so that M_L increases by one. Thus if M_{max} is the maximum M_L of a tableau of SU($2\ell+1$) with spin S and $(-1)^{N_{max}}$ is the phase change of its associated coupling coefficient, then Eq.C-18 gives the phase change of the associated coupling coefficient for any M_L.

$$(-1)^N = (-1)^{N_{max}}(-1)^{M_{max}-M_L} \qquad (C\text{-}18)$$

N_{max} and M_{max} are explicitly given in Eqs. C-19 and C-20 respectively.

$$N_{max} = n^2/2 + nS - n \qquad (C\text{-}19)$$

$$M_{max} = n(2\ +1)/2 - n^2/4 - S^2 \qquad (C\text{-}20)$$

Using Eqs. C-18,C-19, and C-20, we arrive at a general formula (Eq.C-21) for the phase change of the associated coupling coefficient for any M_L.

$$(-1)^N = (-1)^{n(n/4 + S + 1 - \frac{1}{2}) - S^2 - M_L} \qquad (C\text{-}21)$$

Let $|A_n\ LSM_LM_S\rangle$ be a sum of antisymmetric states of SU($4\ell+2$) constructed from the coupling of orbital states $|\ell^n\ L\ M_L\rangle$ and the conjugate spin states $|S\ M_S\rangle$. In Eq.C-22 we have a simple relationship between the associated antisymmetric states of SU($4\ell+2$) in the LHF subshell and the antisymmetric states in the MHF subshell.

$$|(A_n\ LSM_LM_S)^*\rangle = (-1)^N\ |A_{n'}\ LS{-M_L}{-M_S}\rangle \qquad (C\text{-}22)$$

This natural association between the antisymmetric bases of

SU(4ℓ+2) in the LHF and MHF subshells is masked when using the Racah coefficients of fractional parentage to construct these bases. This is clearly evident in the complicated relationship between coefficients of fractional parentage in LHF and MHF configurations.[47]

Using Eqs. C-21 and C-22 we can easily find the matrix elements of double tensor operators in MHF subshells from those of the LHF subshells. We first note that the double tensor operators V^{k1}_{qt} are linear combinations of the operators E_{ij} of SU(4ℓ+2) with t+1=i-j+1 as shown in Eqs. C-24 and C-25.

$$V^{k1}_{qt} = (\ell m \mid V^{k}_{q} \mid \ell m')(\tfrac{1}{2}n \mid V^{1}_{t} \mid \tfrac{1}{2}n')\, E_{mn;m'n'} \qquad \text{(C-24)}$$

$$V^{k\,1}_{-q-t} = (-1)^{q+t}(\ell m \mid V^{k}_{q} \mid \ell m')(\tfrac{1}{2}n \mid V^{1}_{t} \mid \tfrac{1}{2}n')\, E_{m'n';mn} \qquad \text{(C-25)}$$

We use the preceding equations in conjunction with Eqs. C-3, C-21, and C-22 to find the relation (Eq.C-26) between the tensor operators V^{k1}_{qt} and $V^{k\,1}_{-q-t}$.

$$\langle A_n\, LSM_LM_S \mid V^{k1}_{qt} \mid A_n L'S'M_L'M_S' \rangle = \qquad \text{(C-26)}$$

$$(-1)^{S+S'+n+1} \langle A_{n'} LS{-M_L}{-M_S} \mid V^{k\,1}_{-q-t} \mid A_{n'} L'S'{-M_L'}{-M_S'} \rangle$$

Using Eq.C-26 and the definition of the reduced matrix element for a double tensor operator (Eq.C-27), we may now derive the relation (Eq.C-28) between reduced matrix elements of double tensor operators for LHF and MHF subshells.

$$\langle A_n LSM_L M_S | V^{k1}_{qt} | A_n L'S'M'_L M'_S \rangle = (-1)^{L-M_L+S-M_S} \times \tag{C-27}$$

$$\times \begin{pmatrix} L & k & L' \\ -M_L & q & M'_L \end{pmatrix} \begin{pmatrix} S & 1 & S' \\ -M_S & t & M'_S \end{pmatrix} (nLS \| V^{k1} \| nL'S')$$

$$(nLS \| V^{k1} \| nL'S') = (-1)^{L+L'+k} (n'LS \| V^{k1} \| n'L'S') \tag{C-28}$$

Again, if we used the definition Eq.C-14 for the associated orbital bases, we obtain the same result as Racah where the factor $(-1)^{L+L'}$ is unity.

As an example of the use of Eq.C-26 we shall find a simple relation (Eq.C-32) of matrix elements of the spin-orbit operator (Eq.C-29) in MHF subshells to those matrix elements in LHF subshells.

$$\text{S.O.} = \sqrt{\alpha/2}\ (V^{11}_{00} - V^{11}_{-11} - V^{1\,1}_{1-1}) \tag{C-29}$$

Applying Eq.C-26 to the operator S.O. results in Eq.C-30.

$$\langle A_n LSM_L M_S | \text{S.O.} | A_n L'S'M'_L M'_S \rangle = (-1)^{S+S'+n+1} \times \tag{C-30}$$

$$\langle A_{n'} LS{-M_L}{-M_S} | \text{S.O.} | A_{n'} L'S'{-M'_L}{-M'_S} \rangle$$

Finally, using the total angular momentum states (Eq.C-31) and Eq.C-30, we arrive at Eq.C-32.

$$| A_n LS\ JM_J \rangle = \sum_{M_L M_S} (-1)^{L-S+M_J} \sqrt{(2J+1)} \begin{pmatrix} L & S & J \\ M_L & M_S & -M_J \end{pmatrix} | A_n LSM_L M_S \rangle \tag{C-31}$$

$$\langle A_n LS\ JM_J | \text{S.O.} | A_n L'S'\ JM_J \rangle = \tag{C-32}$$

$$(-1)^{L+L'+1} \langle A_{n'} LS\ J{-M_J} | \text{S.O.} | A_{n'} L'S'\ J{-M_J} \rangle$$

Since S.O. commutes with $\hat{J}^2$ we know that the matrix elements in Eq.C-32 are independent of M_J. When L=L' we see that the matrix elements of spin-orbit coupling for MHF subshells are the negative of the matrix elements for LHF subshells. From the selection rules $\Delta L=0,1$ it follows that when $L\neq L'$ the matrix elements are the same in MHF and LHF subshells.

48

APPENDIX D: Eigenvalues of r-Cycle Class Operators

We wish to find the eigenvalues $N_r^p \chi_r^\lambda / \ell^\lambda$ where $[\lambda] = [\lambda_1 \lambda_2 \ldots \lambda_m]$ is a partition labeling the IR's of S_p satisfying Eq. D-1.

$$\sum_{i=1}^{m} \lambda_i = p \qquad (D\text{-}1)$$

By partition we mean that the elements of $[\lambda]$ are monotonically decreasing integers as shown in Eq. D-2.

$$\lambda_1 \geq \lambda_2 \geq \ldots \geq \lambda_m \qquad (D\text{-}2)$$

We first find the $[\lambda]_{ir}$ defined in Eq. D-3 for all i.

$$[\lambda]_{ir} = [\lambda_1 \lambda_2 \ldots \lambda_i - r \ldots \lambda_m] \qquad (D\text{-}3)$$

If $[\lambda]_{ir}$ is not a partition, it may be possible to transform it into one using the following procedure. Let $[R] = [m-1\ m-2 \ldots 0]$ and find the permutation (p_i) such that $(p_i)([\lambda]_{ir} + [R])$ has monotonically decreasing elements. Then find the $[\lambda]'_{ir}$ which are defined in Eq. D-4.

$$[\lambda]'_{ir} = (p_i)([\lambda]_{ir} + [R]) - [R] \qquad (D\text{-}4)$$

Note that $[\lambda]'_{ir}$ is not a partition if and only if $([\lambda]_{ir} + [R])$ contains repeated or negative elements. Also note that $[\lambda]'_{ir} = [\lambda]_{ir}$ if $[\lambda]_{ir}$ is a partition. We shall need only the $[\lambda]'_{ir}$ which are partitions.[49]

We may now find the eigenvalues of the r-cycle class operators by using the simple hook length formula given in Eq. D-5. The sum in this equation is over all i such that $[\lambda]'_{ir}$ is a partition. Also $\epsilon_{(p_i)}$ is 1 or -1 if the permutation (p_i) is an even or odd number of bicycles respectively.

$$N_r^p \chi_r^\lambda / \ell^\lambda = \frac{1}{r} \sum_{i=1}^{m} \epsilon_{(p_i)} \frac{H([\lambda])}{H([\lambda]'_{ir})} \qquad (D\text{-}5)$$

$H([\lambda])$ is the product of hook lengths of partition $[\lambda]$ which has been shown in Fig. 6. As an example, we find the eigenvalue of $\underline{K}_3^9$ for IR $[\lambda] = [432]$ of S_9. The $[\lambda]_{ir}$ are given in Eq. D-6 and the $[\lambda]'_{ir}$ are given in Eq. D-7.

$$\begin{aligned} [432]_{13} &= [132] \\ [432]_{23} &= [402] \\ [432]_{33} &= [43\text{-}1] \end{aligned} \qquad (D\text{-}6)$$

$$\begin{aligned} [432]'_{13} &= (12)\,[342] - [210] = [222] \\ [432]'_{23} &= (23)\,[612] - [210] = [411] \\ [432]'_{33} &= [64\text{-}1] - [210] = [43\text{-}1] \end{aligned} \qquad (D\text{-}7)$$

Using the partitions of Eq. D-7 it is now a simple matter to evaluate the eigenvalue of K_3^9 in terms of hook lengths as shown in Eq. D-8.

$$\frac{N_3^9 \chi_3^{432}}{\ell^{432}} = \frac{1}{3} \left[- \frac{\begin{matrix} 6\ 5\ 3\ 1 \\ 4\ 3\ 1 \\ 2\ 1 \end{matrix}}{\begin{matrix} 4\ 3 \\ 3\ 2 \\ 2\ 1 \end{matrix}} - \frac{\begin{matrix} 6\ 5\ 3\ 1 \\ 4\ 3\ 1 \\ 2\ 1 \end{matrix}}{\begin{matrix} 6\ 3\ 2\ 1 \\ 2 \\ 1 \end{matrix}} \right] \qquad (D\text{-}8)$$

$$= -15$$

TABLE I-a

$$\left\langle \begin{matrix}\frac{1}{2}\\ m'\end{matrix} \right| v^k_q \left| \begin{matrix}\frac{1}{2}\\ m\end{matrix} \right\rangle = (-1)^{\frac{1}{2}-m}\sqrt{2k+1}\begin{pmatrix}\frac{1}{2} & k & \frac{1}{2}\\ -m & q & m\end{pmatrix}$$

$$v^1_{-1} = \begin{bmatrix}\cdot & \cdot\\ 1 & \cdot\end{bmatrix} \qquad v^1_0 = \sqrt{\frac{1}{2}}\begin{bmatrix}1 & \cdot\\ \cdot & -1\end{bmatrix} \qquad v^1_1 = \begin{bmatrix}\cdot & -1\\ \cdot & \cdot\end{bmatrix}$$

$$v^0_0 = \sqrt{\frac{1}{2}}\begin{bmatrix}1 & \cdot\\ \cdot & 1\end{bmatrix}$$

$$m = \quad \tfrac{1}{2} \quad -\tfrac{1}{2}$$

TABLE I-b

$$\left\langle \begin{matrix}1\\ m'\end{matrix} \right| v^k_q \left| \begin{matrix}1\\ m\end{matrix} \right\rangle = (-1)^{1-m}\sqrt{2k+1}\begin{pmatrix}1 & k & 1\\ -m & q & m\end{pmatrix}$$

$$v^2_{-2} = \begin{bmatrix}\cdot & \cdot & \cdot\\ \cdot & \cdot & \cdot\\ 1 & \cdot & \cdot\end{bmatrix} \quad v^2_{-1} = \sqrt{\frac{1}{2}}\begin{bmatrix}\cdot & \cdot & \cdot\\ 1 & \cdot & \cdot\\ \cdot & -1 & \cdot\end{bmatrix} \quad v^2_0 = \sqrt{\frac{1}{6}}\begin{bmatrix}1 & \cdot & \cdot\\ \cdot & -2 & \cdot\\ \cdot & \cdot & 1\end{bmatrix} \quad v^2_1 = \sqrt{\frac{1}{2}}\begin{bmatrix}\cdot & -1 & \cdot\\ \cdot & \cdot & 1\\ \cdot & \cdot & \cdot\end{bmatrix} \quad v^2_2 = \begin{bmatrix}\cdot & \cdot & 1\\ \cdot & \cdot & \cdot\\ \cdot & \cdot & \cdot\end{bmatrix}$$

$$v^1_{-1} = \sqrt{\frac{1}{2}}\begin{bmatrix}\cdot & \cdot & \cdot\\ 1 & \cdot & \cdot\\ \cdot & 1 & \cdot\end{bmatrix} \quad v^1_0 = \sqrt{\frac{1}{2}}\begin{bmatrix}1 & \cdot & \cdot\\ \cdot & 0 & \cdot\\ \cdot & \cdot & -1\end{bmatrix} \quad v^1_1 = \sqrt{\frac{1}{2}}\begin{bmatrix}\cdot & -1 & \cdot\\ \cdot & \cdot & -1\\ \cdot & \cdot & \cdot\end{bmatrix}$$

$$v^0_0 = \sqrt{\frac{1}{3}}\begin{bmatrix}1 & \cdot & \cdot\\ \cdot & 1 & \cdot\\ \cdot & \cdot & 1\end{bmatrix}$$

$$m = \quad 1 \quad 0 \quad -1$$

TABLES II-IV

Unit tensor matrices representing unit tensor operators V_q^k are tabulated according to the convention of Eq. (1). All tensors with different q but with the same k are drawn together into one matrix. The superdiagonal belonging to each q is indicated at the top of each set of tables. The normalization denominator for a superdiagonal is located at its lower end on the right side of each matrix.

II. (j) SUB-SHELL TENSORS

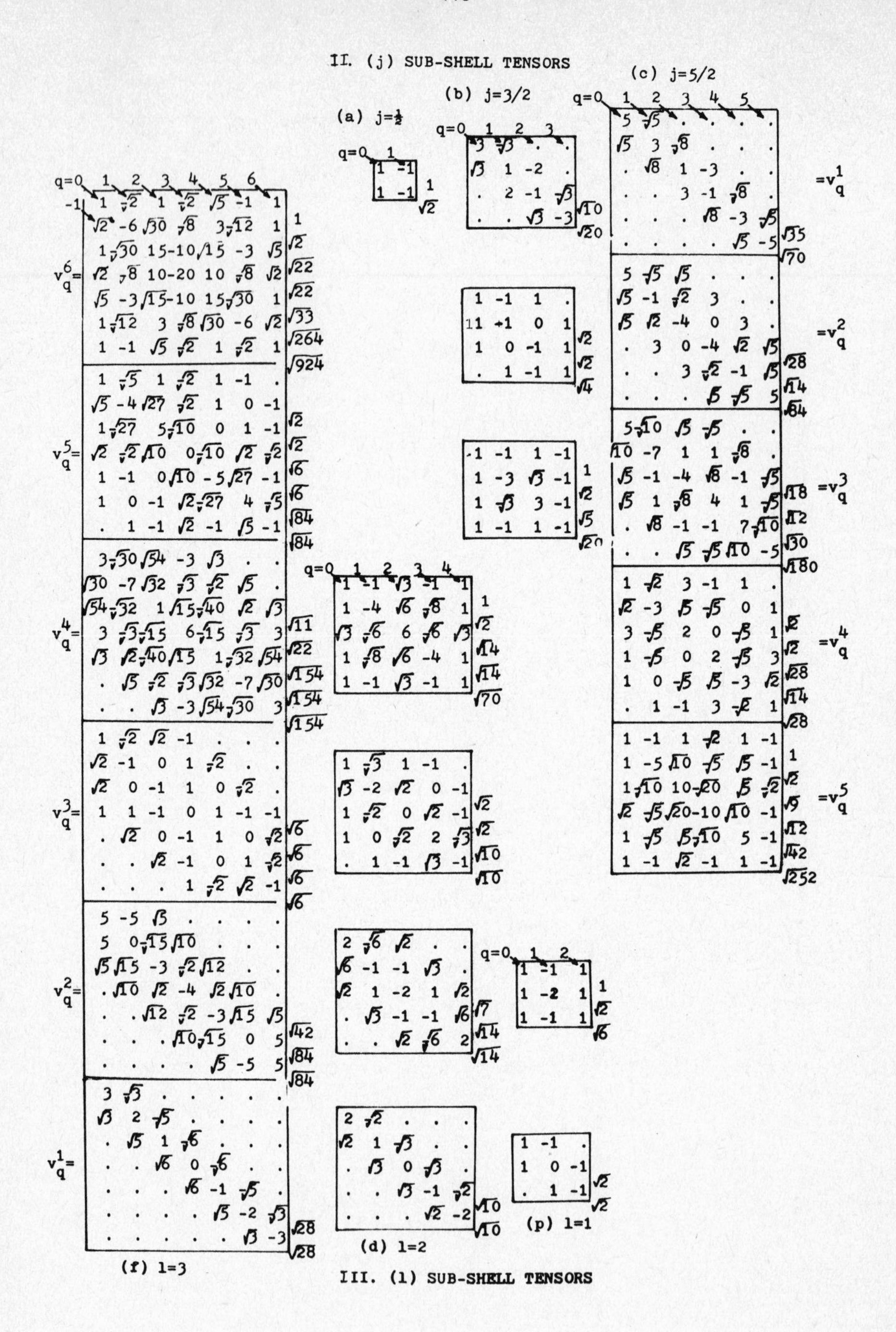

III. (l) SUB-SHELL TENSORS

TABLE III. (g) l=4

V_q^8 =

q=0	1	2	3	4	5	6	7	8	
1	-1	1	-1	$\sqrt{5}$	-1	$\sqrt{7}$	-1	1	1
1	-8	$\sqrt{28}$	-4	$\sqrt{10}$	$-\sqrt{32}$	2	-4	1	$\sqrt{2}$
1	$-\sqrt{28}$	28	-14	$\sqrt{70}$	$-\sqrt{28}$	$\sqrt{56}$	-2	$\sqrt{7}$	$\sqrt{30}$
1	-4	14	-56	$\sqrt{490}$	$-\sqrt{112}$	$\sqrt{28}$	$-\sqrt{32}$	1	$\sqrt{10}$
$\sqrt{5}$	$-\sqrt{10}$	$\sqrt{70}$	$-\sqrt{490}$	70	$-\sqrt{490}$	$\sqrt{70}$	$-\sqrt{10}$	$\sqrt{5}$	$\sqrt{130}$
1	$-\sqrt{32}$	$\sqrt{28}$	$-\sqrt{112}$	$\sqrt{490}$	-56	14	-4	1	$\sqrt{78}$
$\sqrt{7}$	-2	$\sqrt{56}$	$-\sqrt{28}$	$\sqrt{70}$	-14	28	$-\sqrt{28}$	1	$\sqrt{286}$
1	-4	2	$-\sqrt{32}$	$\sqrt{10}$	-4	$\sqrt{28}$	-8	1	$\sqrt{1430}$
1	-1	$\sqrt{7}$	-1	$\sqrt{5}$	-1	1	-1	1	$\sqrt{12870}$

V_q^7 =

q=0	1	2	3	4	5	6	7	8	
1	$-\sqrt{7}$	3	-5	$\sqrt{5}$	-3	1	-1	.	
$\sqrt{7}$	-6	10	-8	$\sqrt{90}$	$-\sqrt{8}$	2	0	-1	$\sqrt{2}$
3	-10	14	$-\sqrt{252}$	$\sqrt{70}$	$-\sqrt{28}$	0	2	-1	$\sqrt{2}$
5	-8	$\sqrt{252}$	-14	$\sqrt{70}$	0	$-\sqrt{28}$	$\sqrt{8}$	-3	$\sqrt{26}$
$\sqrt{5}$	$-\sqrt{90}$	$\sqrt{70}$	$-\sqrt{70}$	0	$\sqrt{70}$	$-\sqrt{70}$	$\sqrt{90}$	$-\sqrt{5}$	$\sqrt{26}$
3	$-\sqrt{8}$	$\sqrt{28}$	0	$-\sqrt{70}$	14	$-\sqrt{252}$	8	-5	$\sqrt{286}$
1	-2	0	$\sqrt{28}$	$-\sqrt{70}$	$\sqrt{252}$	-14	10	-3	$\sqrt{286}$
1	0	-2	$\sqrt{8}$	$-\sqrt{90}$	8	-10	6	$-\sqrt{7}$	$\sqrt{858}$
.	1	-1	3	$-\sqrt{5}$	5	-3	$\sqrt{7}$	-1	$\sqrt{858}$

V_q^6 =

q=0	1	2	3	4	5	6	7	8	
4	$-\sqrt{28}$	2	-4	$\sqrt{40}$	-2	2	.	.	
$\sqrt{28}$	-17	13	-3	$\sqrt{10}$	-1	-1	$\sqrt{7}$	.	
2	-13	22	$-\sqrt{63}$	0	$\sqrt{7}$	$-\sqrt{28}$	1	2	$\sqrt{15}$
4	-3	$\sqrt{63}$	1	$-\sqrt{70}$	$\sqrt{7}$	$-\sqrt{7}$	-1	2	$\sqrt{10}$
$\sqrt{40}$	$-\sqrt{10}$	0	$\sqrt{70}$	-20	$\sqrt{70}$	0	$-\sqrt{10}$	$\sqrt{40}$	$\sqrt{110}$
2	-1	$-\sqrt{7}$	$\sqrt{7}$	$-\sqrt{70}$	1	$\sqrt{63}$	-3	4	$\sqrt{66}$
2	1	$-\sqrt{28}$	$\sqrt{7}$	0	$-\sqrt{63}$	22	-13	2	$\sqrt{33}$
.	$\sqrt{7}$	-1	-1	$\sqrt{10}$	-3	13	-17	$\sqrt{28}$	$\sqrt{660}$
.	.	2	-2	$\sqrt{40}$	-4	2	$-\sqrt{28}$	4	$\sqrt{1980}$

V_q^5 =

q=0	1	2	3	4	5	6	7	8	
4	$-\sqrt{20}$	$\sqrt{20}$	$-\sqrt{80}$	$\sqrt{8}$	-2	.	.	.	
$\sqrt{20}$	-11	$\sqrt{35}$	$-\sqrt{5}$	$-\sqrt{2}$	$\sqrt{5}$	-3	.	.	
$\sqrt{20}$	$-\sqrt{35}$	4	$\sqrt{5}$	$-\sqrt{14}$	$\sqrt{35}$	0	-3	.	
$\sqrt{80}$	$-\sqrt{5}$	$-\sqrt{5}$	9	$-\sqrt{18}$	0	$\sqrt{35}$	$-\sqrt{5}$	-2	$\sqrt{26}$
$\sqrt{8}$	$\sqrt{2}$	$-\sqrt{14}$	$\sqrt{18}$	0	$-\sqrt{18}$	$\sqrt{14}$	$-\sqrt{2}$	$-\sqrt{8}$	$\sqrt{26}$
2	$\sqrt{5}$	$-\sqrt{35}$	0	$\sqrt{18}$	-9	$\sqrt{5}$	$\sqrt{5}$	$-\sqrt{80}$	$\sqrt{234}$
.	3	0	$-\sqrt{35}$	$\sqrt{14}$	$-\sqrt{5}$	-4	$\sqrt{35}$	$-\sqrt{20}$	$\sqrt{78}$
.	.	3	$-\sqrt{5}$	$\sqrt{2}$	$\sqrt{5}$	$-\sqrt{35}$	11	$-\sqrt{20}$	$\sqrt{156}$
.	.	.	2	$-\sqrt{8}$	$\sqrt{80}$	$-\sqrt{20}$	$\sqrt{20}$	-4	$\sqrt{468}$

= V_q^4

q=0	1	2	3	4	5	6	7	8	
14	$-\sqrt{490}$	$\sqrt{630}$	$-\sqrt{70}$	$\sqrt{14}$	.	.	.	.	
$\sqrt{490}$	-21	$\sqrt{70}$	$\sqrt{70}$	$-\sqrt{63}$	$\sqrt{35}$	.	.	.	
$\sqrt{630}$	$-\sqrt{70}$	-11	$\sqrt{360}$	-11	$-\sqrt{10}$	$\sqrt{45}$	.	.	
$\sqrt{70}$	$\sqrt{70}$	$-\sqrt{360}$	9	9	$-\sqrt{360}$	$\sqrt{10}$	$\sqrt{35}$	.	
$\sqrt{14}$	$\sqrt{63}$	-11	-9	18	-9	-11	$\sqrt{63}$	$\sqrt{14}$	$\sqrt{143}$
.	$\sqrt{35}$	$\sqrt{10}$	$-\sqrt{360}$	9	9	$-\sqrt{360}$	$\sqrt{70}$	$\sqrt{70}$	$\sqrt{286}$
.	.	$\sqrt{45}$	$-\sqrt{10}$	-11	$\sqrt{360}$	-11	$-\sqrt{70}$	$\sqrt{630}$	$\sqrt{2002}$
.	.	.	$\sqrt{35}$	$-\sqrt{63}$	$\sqrt{70}$	$\sqrt{70}$	-21	$\sqrt{490}$	$\sqrt{2002}$
.	.	.	.	$\sqrt{14}$	$-\sqrt{70}$	$\sqrt{630}$	$-\sqrt{490}$	14	$\sqrt{2002}$

= V_q^3

q=0	1	2	3	4	5	6	7	8	
14	$-\sqrt{98}$	$\sqrt{14}$	$-\sqrt{14}$	.	.	.	.	.	
$\sqrt{98}$	-7	$-\sqrt{14}$	$\sqrt{14}$	$-\sqrt{35}$	.	.	.	.	
$\sqrt{14}$	$\sqrt{14}$	-13	$\sqrt{8}$	$\sqrt{5}$	$-\sqrt{50}$	.	.	.	
$\sqrt{14}$	$\sqrt{14}$	$-\sqrt{8}$	-9	$\sqrt{45}$	0	$-\sqrt{50}$	.	.	
.	$\sqrt{35}$	$\sqrt{5}$	$-\sqrt{45}$	0	$\sqrt{45}$	$-\sqrt{5}$	$-\sqrt{35}$	.	
.	.	$\sqrt{50}$	0	$-\sqrt{45}$	9	$\sqrt{8}$	$-\sqrt{14}$	$-\sqrt{14}$	$\sqrt{198}$
.	.	.	$\sqrt{50}$	$-\sqrt{5}$	$-\sqrt{8}$	13	$-\sqrt{14}$	$-\sqrt{14}$	$\sqrt{66}$
.	.	.	.	$\sqrt{35}$	$-\sqrt{14}$	$\sqrt{14}$	7	$-\sqrt{98}$	$\sqrt{330}$
.	.	.	.	.	$\sqrt{14}$	$-\sqrt{14}$	$\sqrt{98}$	-14	$\sqrt{990}$

= V_q^2

q=0	1	2	3	4	5	6	7	8	
28	-14	$\sqrt{28}$	.	.	.	.	.	.	
14	7	$-\sqrt{175}$	$\sqrt{63}$	.	.	.	.	.	
$\sqrt{28}$	$\sqrt{175}$	-8	-9	$\sqrt{90}$	.	.	.	.	
.	$\sqrt{63}$	9	-17	$-\sqrt{10}$	10	.	.	.	
.	.	$\sqrt{90}$	$\sqrt{10}$	-20	$\sqrt{10}$	$\sqrt{90}$	.	.	
.	.	.	10	$-\sqrt{10}$	-17	9	$\sqrt{63}$	.	
.	.	.	.	$\sqrt{90}$	-9	-8	$\sqrt{175}$	$\sqrt{28}$	$\sqrt{462}$
.	.	.	.	.	$\sqrt{63}$	$-\sqrt{175}$	7	14	$\sqrt{924}$
.	.	.	.	.	.	$\sqrt{28}$	-14	28	$\sqrt{2772}$

= V_q^1

q=0	1	2	3	4	5	6	7	8	
4	-2	.	.	.	.	.	.	.	
2	3	$-\sqrt{7}$	.	.	.	.	.	.	
.	$\sqrt{7}$	2	-3	.	.	.	.	.	
.	.	3	1	$-\sqrt{10}$	.	.	.	.	
.	.	.	$\sqrt{10}$	0	$-\sqrt{10}$	.	.	.	
.	.	.	.	$\sqrt{10}$	-1	-3	.	.	
.	.	.	.	.	3	-2	$-\sqrt{7}$	.	
.	.	.	.	.	.	$\sqrt{7}$	-3	-2	$\sqrt{60}$
.	.	.	.	.	.	.	2	-4	$\sqrt{60}$

TABLE IV . MIXED SUBSHELL TENSORS

(f) (d)

		2	3	4	5	
	q=1					
	q=0 1	-1	$\sqrt{2}$	$-\sqrt{2}$	1	
	1	$-\sqrt{24}$	3	$-\sqrt{8}$	$\sqrt{3}$	1
	$\sqrt{15}$	$-\sqrt{10}$	$\sqrt{90}$	$-\sqrt{15}$	$\sqrt{5}$	$\sqrt{5}$
$V_q^5 =$	$\sqrt{5}$	$-\sqrt{80}$	$\sqrt{20}$	$-\sqrt{80}$	$\sqrt{5}$	$\sqrt{15}$
	$\sqrt{5}$	$-\sqrt{15}$	$\sqrt{90}$	$-\sqrt{10}$	$\sqrt{15}$	$\sqrt{30}$
	$\sqrt{3}$	$-\sqrt{8}$	3	$-\sqrt{24}$	1	$\sqrt{210}$
	1	$-\sqrt{2}$	$\sqrt{2}$	-1	1	$\sqrt{42}$
	$\sqrt{6}$	$-\sqrt{27}$	3	$-\sqrt{3}$	.	
	$\sqrt{2}$	-7	$\sqrt{48}$	-1	$-\sqrt{2}$	
	$\sqrt{40}$	$-\sqrt{5}$	$\sqrt{15}$	$\sqrt{5}$	$-\sqrt{10}$	$\sqrt{5}$
$V_q^4 =$	$\sqrt{60}$	$-\sqrt{30}$	0	$\sqrt{30}$	$-\sqrt{60}$	$\sqrt{20}$
	$\sqrt{10}$	$-\sqrt{5}$	$-\sqrt{15}$	$\sqrt{5}$	$-\sqrt{40}$	$\sqrt{140}$
	$\sqrt{2}$	1	$-\sqrt{48}$	7	$-\sqrt{2}$	$\sqrt{140}$
	.	$\sqrt{3}$	-3	$\sqrt{27}$	$-\sqrt{6}$	$\sqrt{14}$
	$\sqrt{10}$	$-\sqrt{5}$	$\sqrt{5}$	.	.	
	$\sqrt{10}$	$-\sqrt{15}$	0	$\sqrt{5}$	.	
	$\sqrt{24}$	-1	-3	$\sqrt{3}$	$\sqrt{2}$	
$V_q^3 =$	2	$\sqrt{2}$	$-\sqrt{8}$	$\sqrt{2}$	2	$\sqrt{12}$
	$\sqrt{2}$	$\sqrt{3}$	-3	-1	$\sqrt{24}$	$\sqrt{12}$
	.	$\sqrt{5}$	0	$-\sqrt{15}$	$\sqrt{10}$	$\sqrt{60}$
	.	.	$\sqrt{5}$	$-\sqrt{5}$	$\sqrt{10}$	$\sqrt{30}$
	$\sqrt{5}$	$-\sqrt{5}$	.	.	.	
	$\sqrt{5}$	0	$-\sqrt{5}$	.	.	
	$\sqrt{3}$	$\sqrt{2}$	$-\sqrt{2}$	$-\sqrt{3}$	.	
$V_q^2 =$	1	2	0	-2	-1	
	.	$\sqrt{3}$	$\sqrt{2}$	$-\sqrt{2}$	$-\sqrt{3}$	$\sqrt{14}$
	.	.	$\sqrt{5}$	0	$-\sqrt{5}$	$\sqrt{14}$
	.	.	.	$\sqrt{5}$	$-\sqrt{5}$	$\sqrt{14}$
	$\sqrt{15}$	.	.	.	.	
	$\sqrt{5}$	$\sqrt{10}$	.	.	.	
	1	$\sqrt{8}$	$\sqrt{6}$	.	.	
$V_q^1 =$	.	$\sqrt{3}$	3	$\sqrt{3}$	.	
	.	.	$\sqrt{6}$	$\sqrt{8}$	1	
	.	.	.	$\sqrt{10}$	$\sqrt{5}$	$\sqrt{35}$
	.	.	.	.	$\sqrt{15}$	$\sqrt{35}$

(f) (p)

	3	4	
q=2			
q=1 1	-1	1	
q=0 $\sqrt{3}$	$-\sqrt{12}$	$\sqrt{3}$	1
$\sqrt{3}$	$-\sqrt{15}$	$\sqrt{15}$	2
$\sqrt{10}$	$-\sqrt{8}$	$\sqrt{10}$	$\sqrt{28}$
$\sqrt{15}$	$-\sqrt{15}$	$\sqrt{3}$	$\sqrt{28}$
$\sqrt{3}$	$-\sqrt{12}$	$\sqrt{3}$	$\sqrt{14}$
1	-1	1	
$\sqrt{3}$	$-\sqrt{3}$	.	
$\sqrt{5}$	-2	-1	
1	-1	$-\sqrt{5}$	2
$\sqrt{6}$	0	$-\sqrt{6}$	$\sqrt{12}$
$\sqrt{5}$	1	-1	$\sqrt{12}$
1	2	$-\sqrt{5}$	$\sqrt{2}$
.	$\sqrt{3}$	$-\sqrt{3}$	
$\sqrt{15}$	.	.	
$\sqrt{10}$	$\sqrt{5}$	.	
$\sqrt{2}$	$\sqrt{8}$	1	
$\sqrt{3}$	$\sqrt{3}$	$\sqrt{3}$	$\sqrt{21}$
1	$\sqrt{8}$	$\sqrt{2}$	$\sqrt{21}$
.	$\sqrt{5}$	$\sqrt{10}$	$\sqrt{7}$
.	.	$\sqrt{15}$	

(d) (p)

	2	3	
q=1			
q=0 1	-1	1	
1	$-\sqrt{8}$	$\sqrt{2}$	1
$\sqrt{6}$	$-\sqrt{3}$	$\sqrt{6}$	$\sqrt{3}$
$\sqrt{2}$	$-\sqrt{8}$	1	$\sqrt{15}$
1	-1	1	$\sqrt{5}$

$\sqrt{2}$	$-\sqrt{2}$	.	
1	-1	-1	
$\sqrt{3}$	0	$-\sqrt{3}$	$\sqrt{3}$
1	1	-1	$\sqrt{6}$
.	$\sqrt{2}$	$-\sqrt{2}$	$\sqrt{2}$

$\sqrt{6}$	.	.	
$\sqrt{3}$	$\sqrt{3}$	.	
1	2	1	
.	$\sqrt{3}$	$\sqrt{3}$	$\sqrt{10}$
.	.	$\sqrt{6}$	$\sqrt{10}$

Table V. JJ-Coupling Angular Momentum States

Each square matrix gives the expansion coefficients of the angular momentum states $|J\ M_J\rangle$ to the left in terms of the antisymmetric Gelfand states above. The normalization for each $|J\ M_J\rangle$ is given to the right of the matrix. The numbers given for the expansion coefficients and normalization are the square of their actual value with the appropriate sign retained. For example, -4 is actually $-\sqrt{4}$.

Table VI. JJ-Coupling Tensor Operators

The expansion of the jj-coupling tensor operators W^k_q for even k is shown. All generators are assumed to be added within the parentheses. The normalization is given at the end of the generator expansions and is underlined. The factors in front of the generators and the normalizations are the squares of their actual value with the appropriate sign retained. For example, -4 is actually $-\sqrt{4}$.

TABLE V. JJ-COUPLING ANGULAR MOMENTUM STATES

(j)=3/2

	1 2	
(22)	1	1

	1 3	
(21)	1	1

	1 4	2 3	
(20)	1	1	2
(00)	1	-1	2

(j)=7/2

	1 2	
(66)	1	1

	1 3	
(65)	1	1

	1 4	2 3	
(64)	15	7	22
(44)	7	-15	22

	1 5	2 4	
(63)	4	7	11
(43)	7	-4	11

	1 6	2 5	3 4	
(62)	5	21	7	33
(42)	105	-1	-48	154
(22)	7	-15	20	42

	1 7	2 6	3 5	
(61)	3	28	35	66
(41)	35	15	-27	77
(21)	21	-16	5	42

	1 8	2 7	3 6	4 5	
(60)	1	25	81	25	132
(40)	49	169	-9	-81	308
(20)	49	-1	-9	25	84
(00)	1	-1	1	-1	4

(j)=5/2

	1 2	
(44)	1	1

	1 3	
(43)	1	1

	1 4	2 3	
(42)	9	5	14
(22)	5	-9	14

	1 5	2 4	
(41)	2	5	7
(21)	5	-2	7

	1 6	2 5	3 4	
(40)	1	9	4	14
(20)	25	1	-16	42
(00)	1	-1	1	3

(j)=5/2

	1 2 3	
(9/2 9/2)	1	1

	1 2 4	
(9/2 7/2)	1	1

	1 2 5	1 3 4	
(9/2 5/2)	1	1	2
(5/2 5/2)	1	-1	2

	1 2 6	1 3 5	2 3 4	
(9/2 3/2)	5	32	5	42
(5/2 3/2)	1	0	-1	2
(3/2 3/2)	8	-5	8	21

	1 3 6	1 4 5	2 3 5	
(9/2 1/2)	5	4	5	14
(5/2 1/2)	1	0	-1	2
(3/2 1/2)	1	-5	1	7

TABLE V

(j)=7/2

	1 2 3	
$(\frac{15}{2}\ \frac{15}{2})$	1	1

	1 2 4	
$(\frac{15}{2}\ \frac{13}{2})$	1	1

	1 2 5	1 3 4	
$(\frac{15}{2}\ \frac{11}{2})$	4	3	7
$(\frac{11}{2}\ \frac{11}{2})$	3	-4	7

	1 2 6	1 3 5	2 3 4	
$(\frac{15}{2}\ \frac{9}{2})$	20	64	7	91
$(\frac{11}{2}\ \frac{9}{2})$	45	-4	-28	77
$(\frac{9}{2}\ \frac{9}{2})$	28	-35	80	143

	1 2 7	1 3 6	1 4 5	2 3 5	
$(\frac{15}{2}\ \frac{7}{2})$	5	45	20	21	91
$(\frac{11}{2}\ \frac{7}{2})$	27	12	-3	-35	77
$(\frac{9}{2}\ \frac{7}{2})$	112	-7	-175	135	429
$(\frac{7}{2}\ \frac{7}{2})$	1	-1	1	0	3

	1 2 8	1 3 7	1 4 6	2 3 6	2 4 5	
$(\frac{15}{2}\ \frac{5}{2})$	7	192	375	252	175	1001
$(\frac{11}{2}\ \frac{5}{2})$	21	100	5	-21	-84	231
$(\frac{9}{2}\ \frac{5}{2})$	196	189	-945	361	25	1716
$(\frac{7}{2}\ \frac{5}{2})$	1	0	0	-1	1	3
$(\frac{5}{2}\ \frac{5}{2})$	60	-35	7	15	-15	132

	1 3 8	1 4 7	2 3 7	1 5 6	2 4 6	3 4 5	
$(\frac{15}{2}\ \frac{3}{2})$	35	243	140	100	448	35	1001
$(\frac{11}{2}\ \frac{3}{2})$	112	135	7	5	-140	-63	462
$(\frac{9}{2}\ \frac{3}{2})$	2401	-945	3481	-5040	45	100	12012
$(\frac{7}{2}\ \frac{3}{2})$	1	0	-1	0	0	1	3
$(\frac{5}{2}\ \frac{3}{2})$	125	-189	-5	112	49	-180	660
$(\frac{3}{2}\ \frac{3}{2})$	0	7	-15	-21	12	-15	70

	1 4 8	2 3 8	1 5 7	2 4 7	2 5 6	3 4 6	
$(\frac{15}{2}\ \frac{1}{2})$	28	15	64	175	84	63	429
$(\frac{11}{2}\ \frac{1}{2})$	21	5	12	0	-7	-21	66
$(\frac{9}{2}\ \frac{1}{2})$	245	1701	-2835	605	-540	80	6006
$(\frac{7}{2}\ \frac{1}{2})$	1	0	0	-1	0	1	3
$(\frac{5}{2}\ \frac{1}{2})$	3	35	-21	-27	196	-48	330
$(\frac{3}{2}\ \frac{1}{2})$	49	-105	-28	16	3	-9	210

TABLE V

(j)=7/2

	1 2 3 4	
(88)	1	1

	1 2 3 5	
(87)	1	1

	1 2 3 6	1 2 4 5	
(86)	1	1	2
(66)	1	-1	2

	1 2 3 7	1 2 4 6	1 3 4 5	
(85)	1	5	1	7
(65)	1	0	-1	2
(55)	5	-4	5	14

	1 2 3 8	1 2 4 7	1 2 5 6	1 3 4 6	2 3 4 5	
(84)	7	135	80	135	7	364
(64)	7	15	0	-15	-7	44
(54)	35	3	-64	3	35	140
2(44)	-15	7	0	-7	15	44
4(44)	15	-7	21	-7	15	65

	1 2 4 8	1 2 5 7	1 3 4 7	1 3 5 6	2 3 4 6	
(83)	7	25	27	25	7	91
(63)	7	4	0	-4	-7	22
(53)	21	-12	4	-12	21	70
2(43)	-4	7	0	-7	4	22
4(43)	16	7	-84	7	16	130

	1 2 5 8	1 2 6 7	1 3 4 8	1 3 5 7	2 3 4 7	1 4 5 6	2 3 5 6	
(82)	189	125	175	1024	175	125	189	2002
(62)	21	5	7	0	-7	-5	-21	66
(52)	21	-45	112	-64	112	-45	21	420
2(42)	-1	105	-48	0	48	-105	1	308
4(42)	529	105	-108	-336	-108	105	529	1820
2(22)	15	-7	-20	0	20	7	-15	84
4(22)	15	-63	-20	35	-20	-63	15	231

	1 2 6 8	1 3 5 8	1 3 6 7	2 3 4 8	1 4 5 7	2 3 5 7	2 4 5 6	
(81)	7	35	27	5	27	35	7	143
(61)	28	35	3	0	-3	-35	-28	132
(51)	0	7	-15	16	-15	7	0	60
2(41)	15	-27	35	0	-35	27	-15	154
4(41)	375	-3	-35	-84	-35	-3	375	910
2(21)	16	-5	-21	0	21	5	-16	84
4(21)	-36	125	-21	-560	-21	125	-36	924

	1 2 7 8	1 3 6 8	1 4 5 8	1 4 6 7	2 3 5 8	2 3 6 7	2 4 5 7	3 4 5 6	
(80)	7	175	112	135	135	112	175	7	858
(60)	4	25	4	0	0	-4	-25	-4	66
(50)	0	0	0	-1	1	0	0	0	2
2(40)	9	4	-64	0	0	64	-4	-9	154
4(40)	45	20	-5	-21	-21	-5	20	45	182
2(20)	16	-4	1	0	0	-1	4	-16	42
4(20)	-12	3	27	-35	-35	27	3	-12	77
(00)	-1	1	-1	0	0	-1	1	-1	6

TABLE VI. JJ-COUPLING TENSOR OPERATORS

(p) Sub-shell

$$W_0^2 = (E_{11} \quad -E_{22} \quad 2E_{52} \quad -E_{33} \quad -2E_{63} \quad E_{44} \quad 2E_{25} \quad -2E_{36})/ \underline{6}$$

$$-W_1^2 = (2E_{12} \quad 3E_{53} \quad -2E_{34} \quad -E_{64} \quad -E_{15} \quad 3E_{26})/ \underline{6}$$

$$W_2^2 = (E_{13} \quad E_{24} \quad 2E_{54} \quad -2E_{16})/ \underline{3}$$

(d) Sub-shell

$$W_0^2 = (100E_{11} \quad -4E_{22} \quad 36E_{72} \quad -64E_{33} \quad 6E_{83} \quad -64E_{44} \quad -6E_{94} \quad -4E_{55} \quad -36E_{105} \quad 100E_{66} \quad 36E_{27} \quad 49E_{77} \quad 6E_{38} \quad -49E_{88} \quad -6E_{49} \quad -49E_{99} \quad -36E_{510} \quad 49E_{1010})/ \underline{350}$$

$$-W_1^2 = (120E_{12} \quad 48E_{23} \quad 27E_{73} \quad 25E_{84} \quad -48E_{45} \quad 2E_{95} \quad -120E_{56} \quad -30E_{106} \quad -30E_{17} \quad 2E_{28} \quad 98E_{78} \quad 25E_{39} \quad 27E_{410} \quad -98E_{910})/ \underline{350}$$

$$W_2^2 = (30E_{13} \quad 54E_{24} \quad 6E_{74} \quad 54E_{35} \quad 16E_{85} \quad 30E_{46} \quad 20E_{96} \quad -20E_{18} \quad -16E_{29} \quad 49E_{79} \quad -6E_{310} \quad 49E_{810})/ \underline{175}$$

$$W_0^4 = (E_{11} \quad -9E_{22} \quad 4E_{72} \quad 4E_{33} \quad -24E_{83} \quad 4E_{44} \quad 24E_{94} \quad -9E_{55} \quad -4E_{105} \quad E_{66} \quad 4E_{27} \quad -24E_{38} \quad 24E_{49} \quad -4E_{510})/ \underline{70}$$

$$-W_1^4 = (4E_{12} \quad -10E_{23} \quad 10E_{73} \quad -30E_{84} \quad 10E_{45} \quad 15E_{95} \quad -4E_{56} \quad -E_{106} \quad -E_{17} \quad 15E_{28} \quad -30E_{39} \quad 10E_{410})/ \underline{70}$$

$$W_2^4 = (9E_{13} \quad -5E_{24} \quad 20E_{74} \quad -5E_{35} \quad -30E_{85} \quad 9E_{46} \quad 6E_{96} \quad -6E_{18} \quad 30E_{29} \quad -20E_{310})/ \underline{70}$$

$$-W_3^4 = (2E_{14} \quad 5E_{75} \quad -2E_{36} \quad -3E_{86} \quad -3E_{19} \quad 5E_{210})/ \underline{10}$$

$$W_4^4 = (E_{15} \quad E_{26} \quad 4E_{76} \quad -4E_{110})/ \underline{5}$$

TABLE VI. JJ-COUPLING TENSOR OPERATORS

(f) Sub-shell

$$
\begin{array}{lllllll}
W_0^2 = (& 1225E_{11} & 25E_{22} & 150E_{92} & -225E_{33} & 90E_{10\,3} & -625E_{44} \\
 & 12E_{11,4} & -625E_{55} & -12E_{12,5} & -225E_{66} & -90E_{13,6} & 25E_{77} \\
 & -150E_{14,7} & 1225E_{88} & 150E_{29} & 900E_{99} & 90E_{3,10} & -36E_{1010} \\
 & 12E_{4,11} & -576E_{1111} & -12E_{5,12} & -576E_{1212} & -90E_{6,13} & -36E_{1313} \\
 & -150E_{7,14} & 900E_{1414})/\ \underline{4116} & & & &
\end{array}
$$

$$
\begin{array}{lllllll}
-W_1^2 = (& 1050E_{12} & 800E_{23} & 75E_{93} & 250E_{34} & 121E_{10,4} & 98E_{11,5} \\
 & -250E_{56} & 30E_{12,6} & -800E_{67} & -5E_{13,7} & -1050E_{78} & -175E_{14,8} \\
 & -175E_{19} & -5E_{2,10} & 1080E_{9,10} & 30E_{3,11} & 432E_{1011} & 98E_{4,12} \\
 & 121E_{5,13} & -432E_{1213} & 75E_{6,14} & -1080E_{1314})/\ \underline{4116} & &
\end{array}
$$

$$
\begin{array}{lllllll}
W_2^2 = (& 175E_{13} & 375E_{24} & 10E_{94} & 500E_{35} & 32E_{10,5} & 500E_{46} \\
 & 60E_{11,6} & 375E_{57} & 80E_{12,7} & 175E_{68} & 70E_{13,8} & -70E_{1,10} \\
 & -80E_{2,11} & 270E_{9,11} & -60E_{3,12} & 486E_{1012} & -32E_{4,13} & 486E_{1113} \\
 & -10E_{5,14} & 270E_{1214})/\ \underline{2058} & & & &
\end{array}
$$

$$
\begin{array}{lllllll}
W_0^4 = (& 441E_{11} & -1521E_{22} & 600E_{92} & -81E_{33} & -640E_{10,3} & 729E_{44} \\
 & -300E_{11,4} & 729E_{55} & 300E_{12,5} & -81E_{66} & 640E_{13,6} & -1521E_{77} \\
 & -600E_{14,7} & 441E_{88} & 600E_{29} & 121E_{99} & -640E_{3,10} & -1089E_{1010} \\
 & -300E_{4,11} & 484E_{1111} & 300E_{5,12} & 484E_{1212} & 640E_{6,13} & -1089E_{1313} \\
 & -600E_{7,14} & 121E_{1414})/\ \underline{7546} & & & &
\end{array}
$$

$$
\begin{array}{lllllll}
-W_1^4 = (& 1260E_{12} & -540E_{23} & 1000E_{93} & -972E_{34} & -120E_{10,4} & -735E_{11,5} \\
 & 972E_{56} & -E_{12,6} & 540E_{67} & 1014E_{13,7} & -1260E_{78} & -210E_{14,8} \\
 & -210E_{19} & 1014E_{2,10} & 484E_{9,10} & -E_{3,11} & -1210E_{1011} & -735E_{4,12} \\
 & -120E_{5,13} & 1210E_{1213} & 1000E_{6,14} & -484E_{1314})/\ \underline{7546} & &
\end{array}
$$

$$
\begin{array}{lllllll}
W_2^4 = (& 1890E_{13} & 18E_{24} & 1200E_{94} & -864E_{35} & 60E_{10,5} & -864E_{46} \\
 & -578E_{11,6} & 18E_{57} & -486E_{12,7} & 1890E_{68} & 756E_{13,8} & -756E_{1,10} \\
 & 486E_{2,11} & 1089E_{9,11} & 578E_{3,12} & -605E_{1012} & -60E_{4,13} & -605E_{1113} \\
 & -1200E_{5,14} & 1089E_{1214})/\ \underline{7546} & & & &
\end{array}
$$

$$
\begin{array}{lllllll}
-W_3^4 = (& 252E_{14} & 144E_{25} & 150E_{95} & 98E_{10,6} & -144E_{47} & -3E_{11,7} \\
 & -252E_{58} & -189E_{12,8} & -189E1{,}11 & -3E_{2,12} & 242E_{9,12} & 98E_{3,13} \\
 & 150E_{4,14} & -242E_{1114})/\ \underline{1078} & & & &
\end{array}
$$

$$
\begin{array}{lllllll}
W_4^4 = (& 63E_{15} & 135E_{26} & 40E_{96} & 135\ E_{37} & 96E_{10,7} & 63E_{48} \\
 & 84E_{11,8} & -84E_{1,12} & -96E_{2,13} & 121\ E_{9,13} & -40E_{3,14} & 121E_{1014})/\ \underline{539}
\end{array}
$$

TABLE VI JJ-COUPLING TENSOR OPERATORS

(f) Sub-shell

W^6_0 = (E_{11} $-25E_{22}$ $6E_{92}$ $81E_{33}$ $-90E_{10,3}$ $-25E_{44}$
$300E_{11,4}$ $-25E_{55}$ $-300E_{12,5}$ $81E_{66}$ $90E_{13,6}$ $-25E_{77}$
$-6E_{14,7}$ E_{88} $6E_{29}$ $-90E_{3,10}$ $300E_{4,11}$ $-300E_{5,12}$
$90E_{6,13}$ $-6E_{7,14}$)/ $\underline{924}$

$-W^6_1$ = ($6E_{12}$ $-56E_{23}$ $21E_{93}$ $70E_{34}$ $-175E_{10,4}$ $350E_{11,5}$
$-70E_{56}$ $-210E_{12,6}$ $56E_{67}$ $35E_{13,7}$ $-6E_{78}$ $-E_{14,8}$
$-E_{19}$ $35E_{2,10}$ $-210E_{3,11}$ $350E_{4,12}$ $-175E_{5,13}$ $21E_{6,14}$)/ $\underline{924}$

W^6_2 = ($5E_{13}$ $-21E_{24}$ $14E_{94}$ $7E_{35}$ $-70E_{10,5}$ $7E_{46}$
$84E_{11,6}$ $-21E_{57}$ $-28E_{12,7}$ $5E_{68}$ $2E_{13,8}$ $-2E_{1,10}$
$28E_{2,11}$ $-84E_{3,12}$ $70E_{4,13}$ $-14E_{5,14}$)/ $\underline{231}$

$-W^6_3$ = ($8E_{14}$ $-14E_{25}$ $21E_{95}$ $-63E_{10,6}$ $14E_{47}$ $42E_{11,7}$
$-8E_{58}$ $-6E_{12,8}$ $-6E_{1,11}$ $42E_{2,12}$ $-63E_{3,13}$ $21E_{4,14}$)/ $\underline{154}$

W^6_4 = ($15E_{15}$ $-7E_{26}$ $42E_{96}$ $-7E_{37}$ $-70E_{10,7}$ $15E_{48}$
$20E_{11,8}$ $-20E_{1,12}$ $70E_{2,13}$ $-42E_{3,14}$)/ $\underline{154}$

$-W^6_5$ = ($2E_{16}$ $7E_{97}$ $-2E_{38}$ $-5E_{10,8}$ $-5E_{1,13}$ $7E_{2,14}$)/ $\underline{14}$

W^6_6 = (E_{17} E_{28} $6E_{98}$ $-6E_{1,14}$)/ $\underline{7}$

REFERENCES

1. W. G. Harter, Phys. Rev. A 8 6 2819 (1973).

2. G. Racah, Phys. Rev. 76 769 (1949).

3. I. M. Gelfand and M. Zetlin, Dokl. Akad. Nauk. SSSR 71 825 (1950)

4. G. E. Baird and L C. Biedenharn, J. Math. Phys. 4 1449 (1963).

5. L. C. Biedenharn, A. Giovannini, and J. D. Louck, J. Math. Phys. 8 691 (1967).

6. J. D. Louck, Am. J. Phys. 38 3 (1970).

7. M. Moshinsky, in Physics of Many Particle Systems, edited by E. Meeron (Gordan and Breech, New York, 1966).

8 D. E. Rutherford, Substitutional Analysis Edinburgh (1947).

9. H. Weyl, The Classic Groups (Princeton Univ. Press,1946).

10. H. Weyl, Theory of Groups and Quantum Mechanics (Dover Publications, 1931).

11. T. Yamanouchi, Proc. Phys.-Math. Soc. Japan 19 436 (1937).

12. H. A. Jahn and H. vanWieringen, Proc. Roy. Soc. (London) A 209 502 (1951).

13. G. deB. Robinson, Representation Theory of Symmetric Groups, (University of Toronto Press, Toronto Canada, 1961).

14. W. A. Goddard III, Phys. Rev. 157 1 73 (1967).

15. Rotenburg, et. al., 3-j and 6-j Symbols, (MIT Press, Cambridge Mass. 1959).

16. C. W. Nielson, G. F. Koster, Spectroscopic Coefficients for the p^n d^n and f^n Configurations, (MIT Press, Cambridge Mass. 1963).

17. W.G. Harter, J. Math. Phys. (to be published).

18. Ref. 13.

19. A. J. Coleman, Induced Representations with Applications to S_n, (Queens Univ., Kingston, Ont. Canada, 1966).

20. M. Ciftan and L. C. Biedenharn, J. Math. Phys. 10, 221 (1969).

21. Ref. 11.

22. Ref. 4-6.

23. Ref. 9-14.

24. E. Wigner, Group Theory, (Academic Press, New York 1959), p. 151.

25. Ref. 4-6.

26. C. W. Patterson, Thesis submitted in partial fulfillment of requirements for degree of Doctor of Philosophy at University of Southern California, August 1974.

27. Ref. 13 and 22.

28. J. H. Staib, An Introduction to Matrices and Linear Transformations (Addison-Wesley, Reading, Mass., 1969), p. 259.

29. J. Caird (in preparation).

30. W. G. Harter, Phys. Rev. a 8, 2819 (1973).

31. Ref. 15, Eq. (1.21).

32. Ref. 4, Eq. (60).

33. R. Paunz, Alternant Molecular Orbital Method (W. B. Saunders, Philadelphia, Pa., 1967).

34. W. Heitler and F. London, Z. Phys. 44, 455 (1927).

35. H. A. Jahn and E. Teller, Proc. Roy. Soc. (London) A161 220 (1937).

36. I. M. Gelfand, Mat. Sb. 26, 103 (1950).

37. A. M. Perelomov and V. S. Popov, J. Nucl. Phys. (USSR) 3, 924 (1966), translated in Soviet JNP 3, 676 (1966).

38. K. J. Lezuo, J. Math. Phys. 13, 1389 (1972).

39. A. Partensky, J. Math. Phys. 13, 1503 (1972).

40. Ref. 26, p. 104.

41. Ref. 39, Eq. (16).

42. Ref. 29, Eqs. 4-23(a,c), pp. 45-46.

43. Ref. 6, Eqs. (2-33)-(2-38), p. 8.

44. G. E. Baird and L. C. Biedinharn, J. Math. Phys. 5, 1723 (1964).

45. Ref. 44, Eq. (10).

46. G. Racah, Phys. Rev. 63, 367 (1943).

47. Ref. 46.

48. Ref. 19. The results given here are based on this work.

49. This treatment is remarkably similar to that used to find the Clebsch-Gordon series for U_n. See J. D. Louck, Am. J. Phys. 38, 18 (1970).

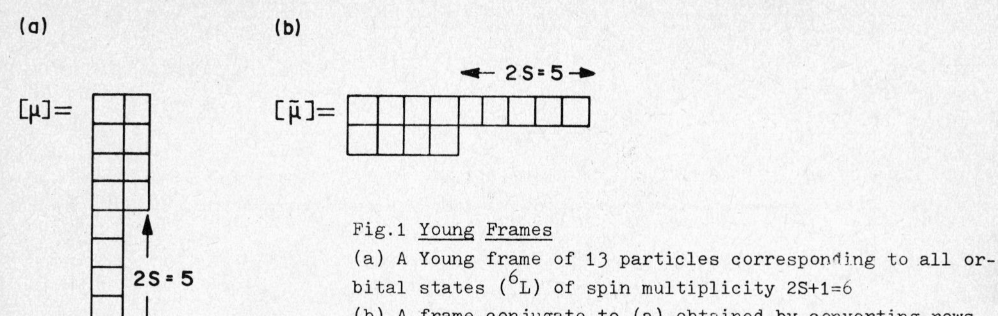

Fig.1 <u>Young Frames</u>

(a) A Young frame of 13 particles corresponding to all orbital states (^6L) of spin multiplicity 2S+1=6

(b) A frame conjugate to (a) obtained by converting rows to columns, corresponds to spin states of total spin S=5/2, since only 5 of the 13 spins are unpaired. (These are represented by the single row of 5 boxes.)

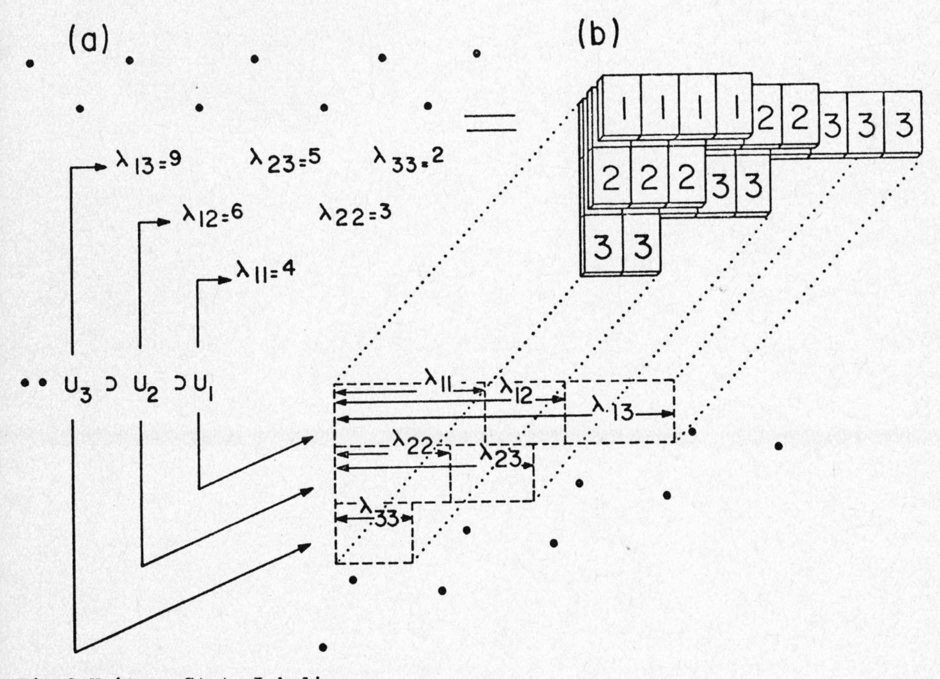

Fig.2 Unitary State Labeling

(a) Gelfand Pattern - The jth row of integers ($\lambda_{1,j}\lambda_{2,j}\cdots$ $..\lambda_{j,j}$) tells to which representation of U_j the state belongs, and similarily for the j-1th row ($\lambda_{1,j-1}\lambda_{2,j-i}\cdots\lambda_{j-1,j-1}$) which labels a unique representation of U_{j-1} contained in ($\lambda_{1,j}\lambda_{2,j}\cdots\lambda_{j,j}$). In this way each state has a unique genealogy chain and labeling.

(b) Young Tableau - Tableaus are a completely equivalent but non-algebraic "picture" of the Gelfand patterns. (When labeled algebraically, it is just an up-side-down Gelfand Pattern.)

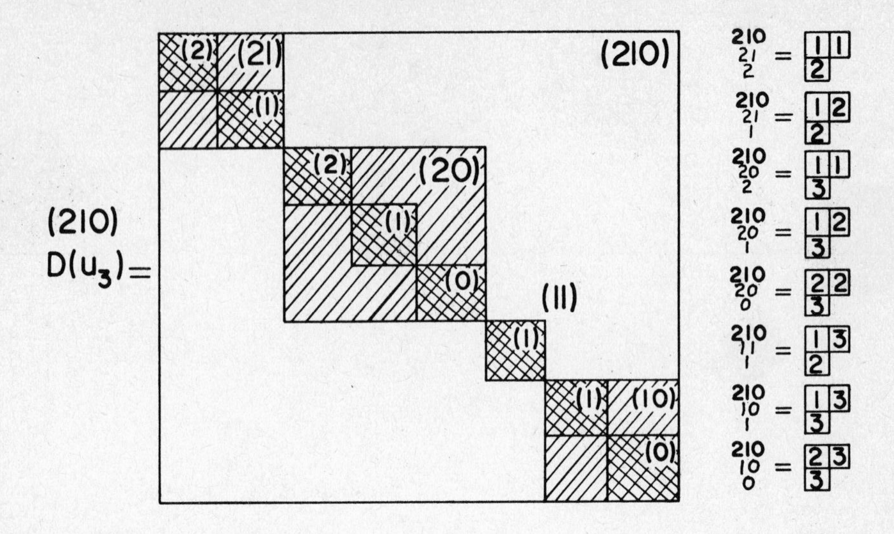

Fig.3 Form of Gelfand Representation
A Gelfand representation (210) of U_3 is irreducible, but as a representation of subgroup U_2 it is reduced to (21) ⊕ (20) ⊕ (11) ⊕ (10) as shown. As a representation of U_1 it is a diagonal, and each base state corresponds to a diagonal component with a unique "address" or "genealogy" traced by a Gelfand pattern or tableau on the right.

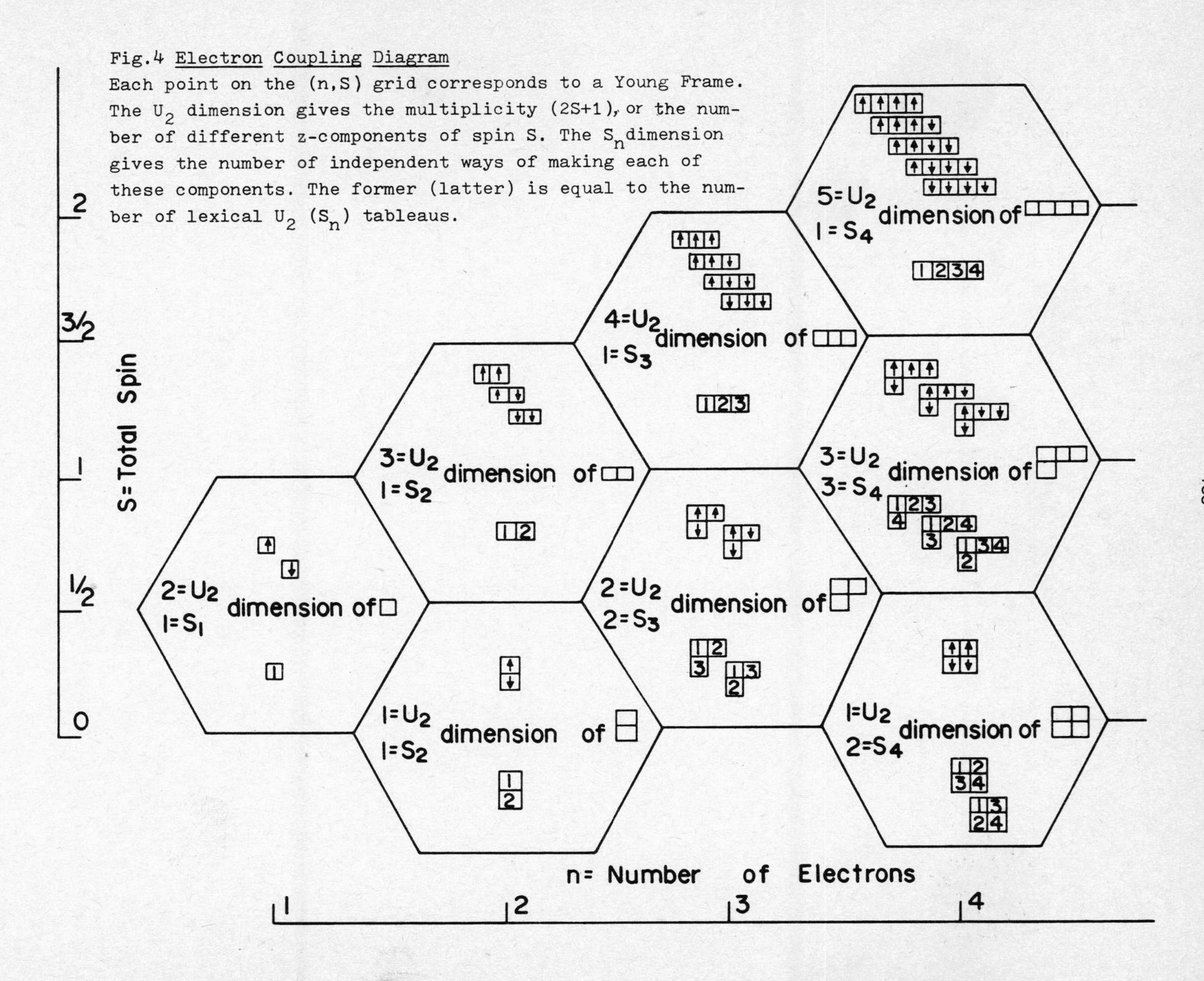

Fig.4 Electron Coupling Diagram

Each point on the (n,S) grid corresponds to a Young Frame. The U_2 dimension gives the multiplicity (2S+1), or the number of different z-components of spin S. The S_n dimension gives the number of independent ways of making each of these components. The former (latter) is equal to the number of lexical U_2 (S_n) tableaus.

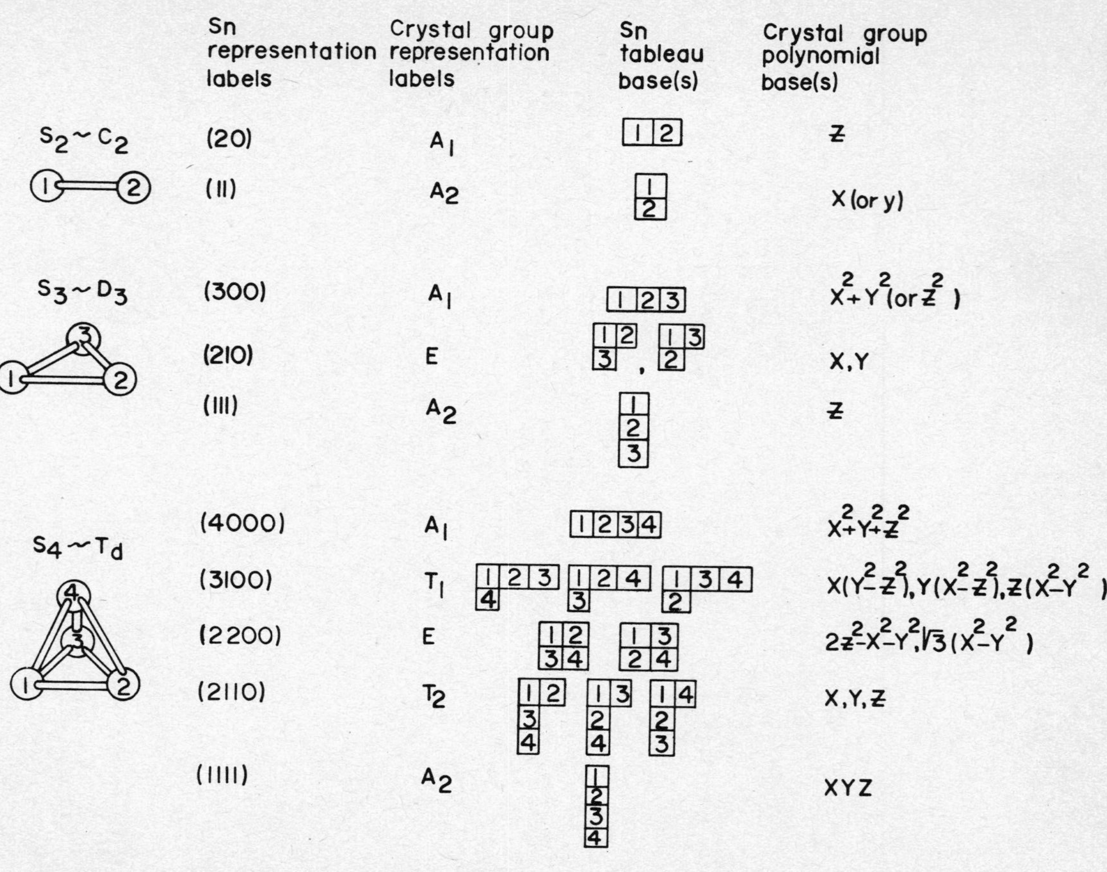

Fig. 5 Correspondence Between S_n and Point Group Bases
Each permutation of numbers corresponds to a single point group operation in each case. Each tableau corresponds to a point group base polynomial.

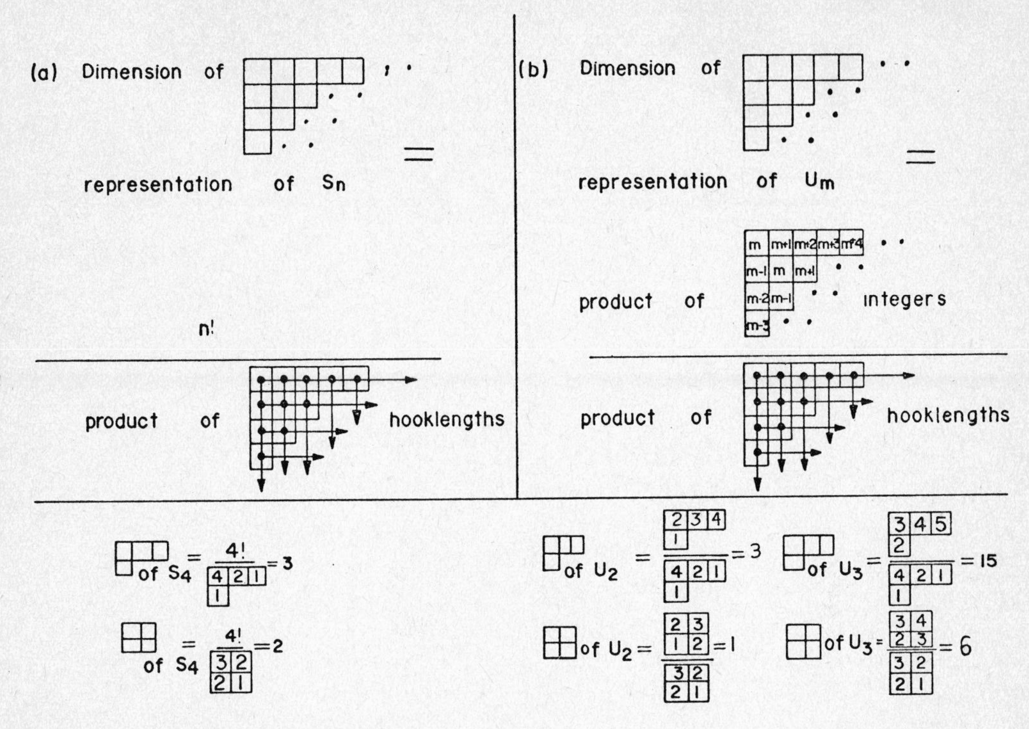

Fig.6 Hall - Robinson Hooklength Formulas
Dimension of representations of (a) S_n and (b) U_m labeled by a single tableau are given by the formulas. A hooklength of a tableau box is simply the number of boxes in a "hook" consisting of all the boxes below it, to the right of it, and itself.

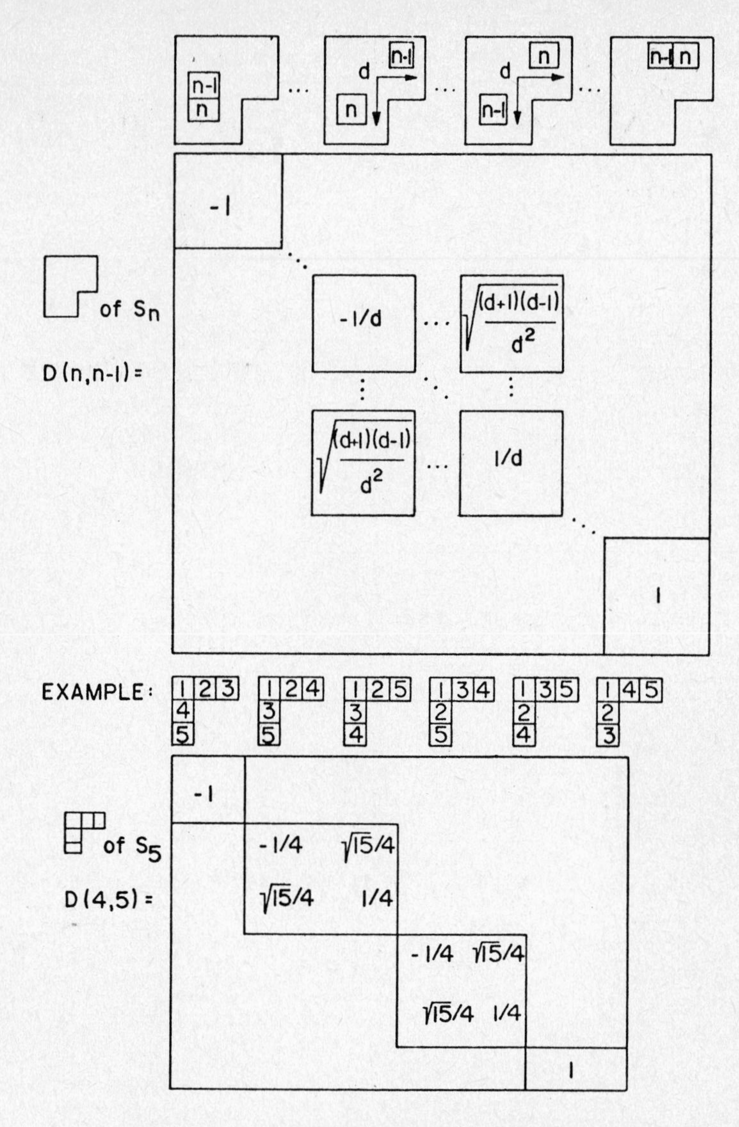

Fig.7 Yamanouchi Formulas for Permutation Operators
As explained in the text, d is the "city block" distance between the relevant blocks, i.e. the minimum number of streets to be crossed when traveling from one to the other. Note that when numbers (n) and (n-1) are ordered smaller above larger, the permutation is negative (anti-symmetric if d=1), and positive (symmetric if d=1)when the smaller number is left of the larger number.(The (n-1) will never be above and left of (n) since that arrangement would be "non-lexical.")

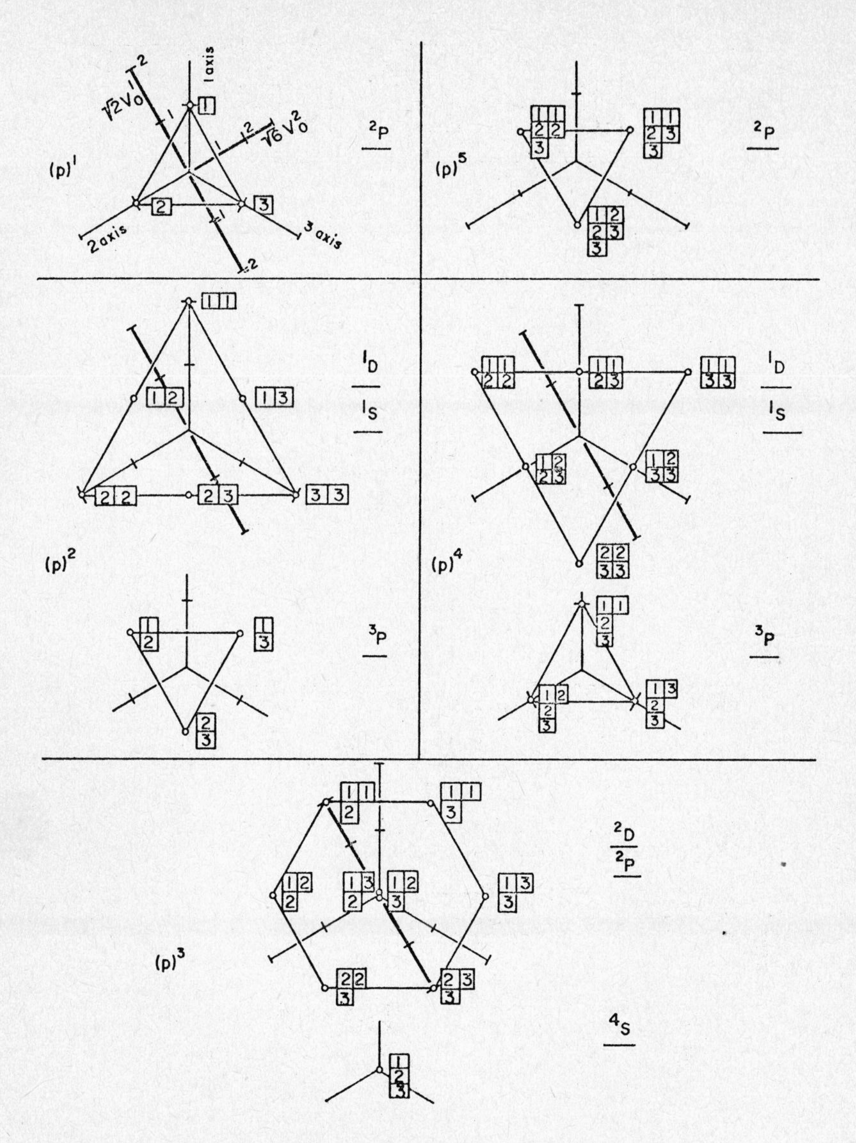

Fig.8 Weight or Moment Diagrams of Atomic $(p)^n$ States
Each tableau is located at point $(x_1\ x_2\ x_3)$ in a cartesian co-ordinate system for which x_n is the number of n's in the tableau. An alternative co-ordinate system is $(v_0^2,\ v_0^1,\ v_0^0)$ defined by Eq.16 which gives the zz-quadrupole moment, z-magnetic dipole moment, and number of particles, respectively. The last axis (v_0^0) would be pointing straight out of the figure, and each family of states lies in a plane perpendicular to it.

(a) $\langle T' | E_{ii} | T \rangle = \delta_{T',T} \begin{pmatrix} \text{number} \\ \text{of (i)'s} \end{pmatrix}$

(b) $\langle T' | E_{ij} | T \rangle = \langle T | E_{ji} | T' \rangle$

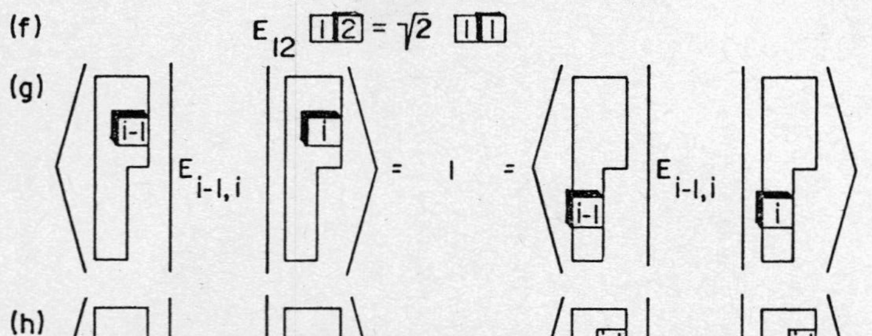

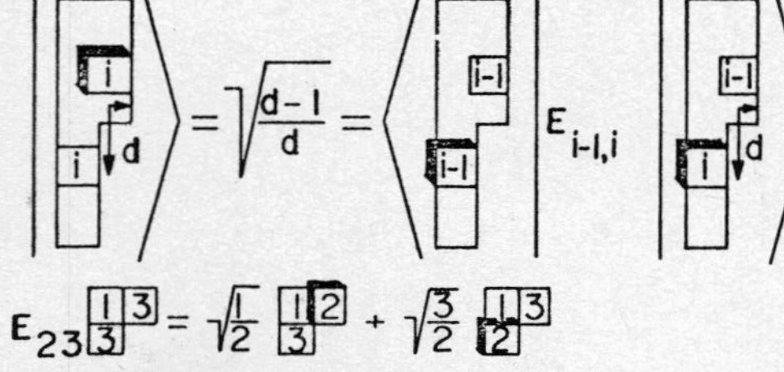

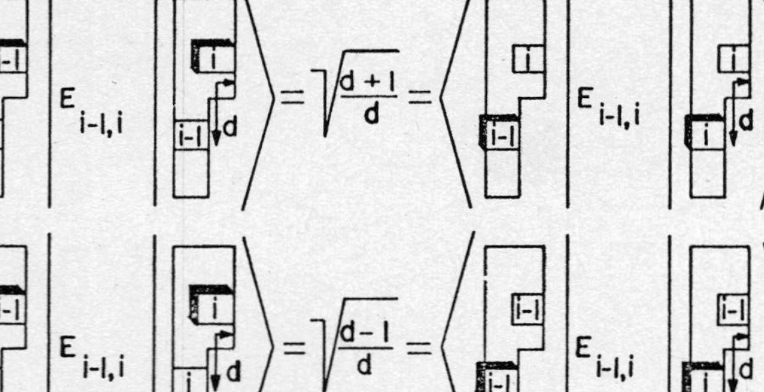

(e) E_{23} [1 3 / 3] $= \sqrt{\frac{1}{2}}$ [1 2 / 3] $+ \sqrt{\frac{3}{2}}$ [1 3 / 2]

(f) E_{12} [1 2] $= \sqrt{2}$ [1 1]

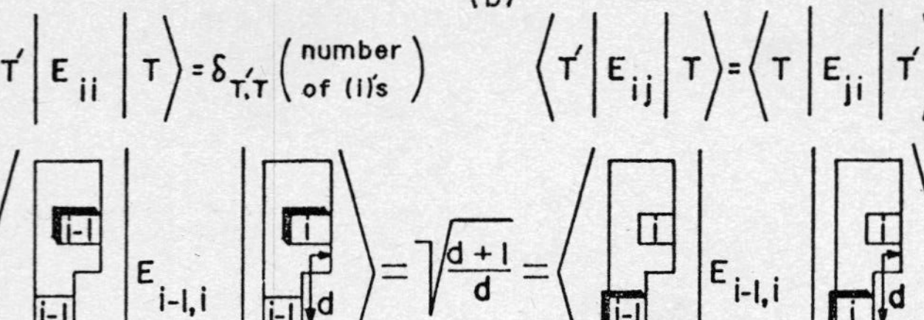

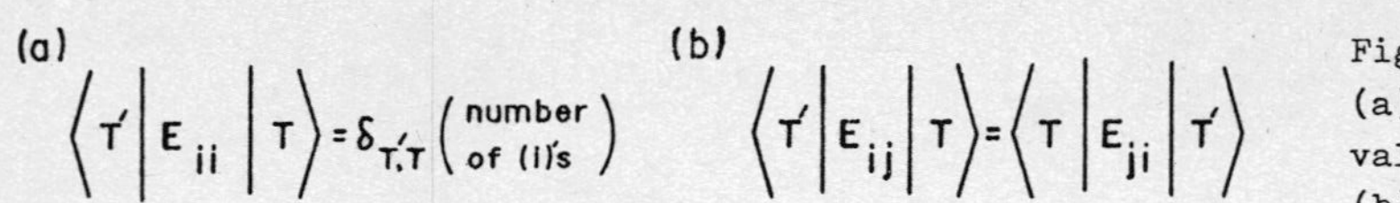

Fig.9 Tableau Formulas for Electronic Orbital Operators
(a) Number operators E_{ii} are diagonal.(The only eigenvalues for orbitalstates are 0,1,and 2.)
(b) Raising and lowering operators are simply transposes of each other.
(c-h) $E_{i-1,i}$ acting on a tableau state gives zero unless there is an (i) in a column of the tableau that doesn't already have an (i-1),too. Then it gives back a new state with the (i) changed to(i-1) and a factor (matrix element) that depends on where the other (i)'s and (i-1)'s are located. (Boxes not outlined in the figure contain numbers not equal to (i) or (i-1).) Cases (c) and (d) involved the "city block" distance d (See Fig.7) which is the denominator of the matrix element. The numerator is one larger (d+1) or smaller (d-1) depending on whether the involved tableaus favor the larger or smaller state number (i or i-1) with a higher position. The special cases of (d=1) shown in (f) always pick the larger (and non-zero) choice of d+1=2. All other non-zero matrix elements are equal to unity.

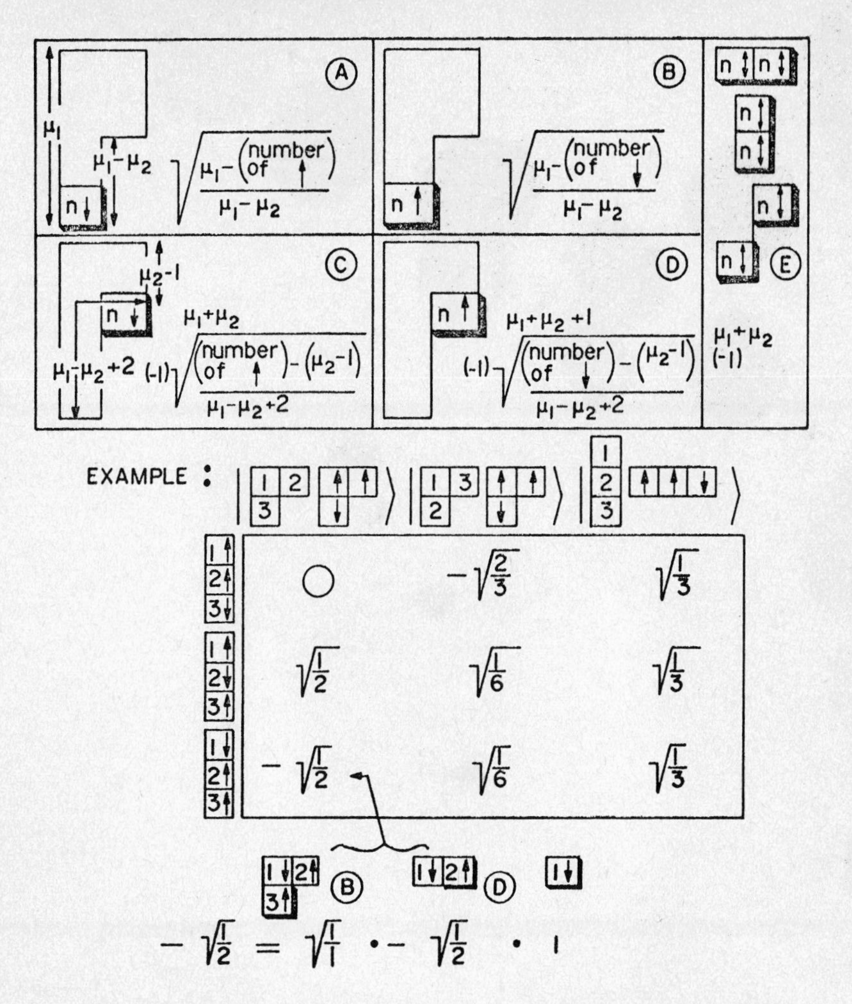

Fig.10 Tableau Formulas for Combining Orbital and Spin States
The five formulas above (A-E) give the possible factors for the coefficient of transformation between a Slater state (verticle tableau) and orbital-spin state. We first note the spin associated with every orbital number in the Slater state and write these spins within the orbital state tableau. We then proceed to rip off boxes with numbered spins in the orbital state tableau starting with the highest orbital number. Each "rip-off" gives a factor depending on the position and associated spin of the orbital number (Cases A-E). The product of successive factors gives the coefficient desired. All numbers in the formulas refer to condition of tableau just before the box outlined in the figure is removed.

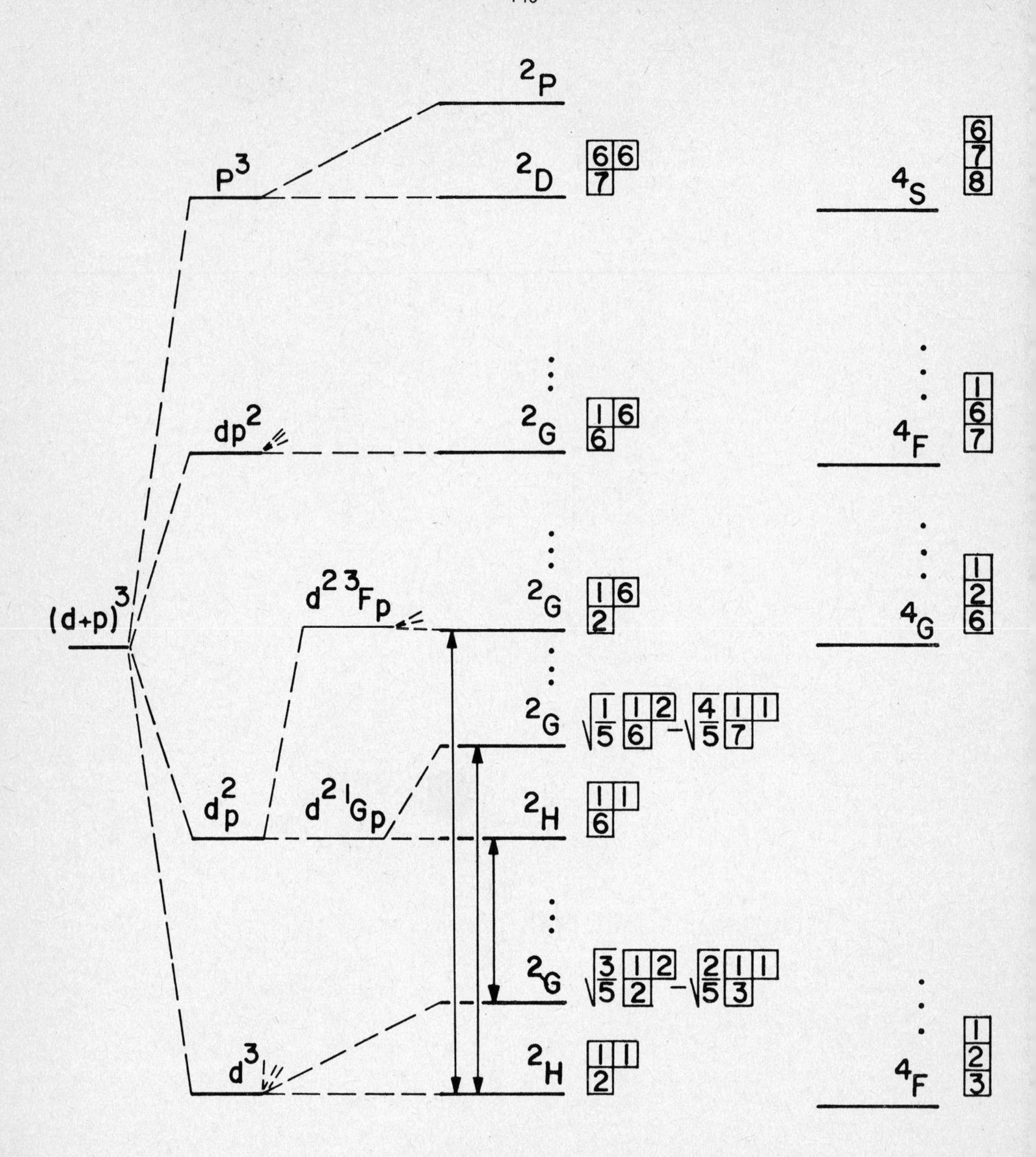

Fig. 11 Example of Unitary Tableau Notation for Multiple Shell States.
The calculation of the dipole operator using graphical formula between states of definite spin and orbit can be done as explained in text.

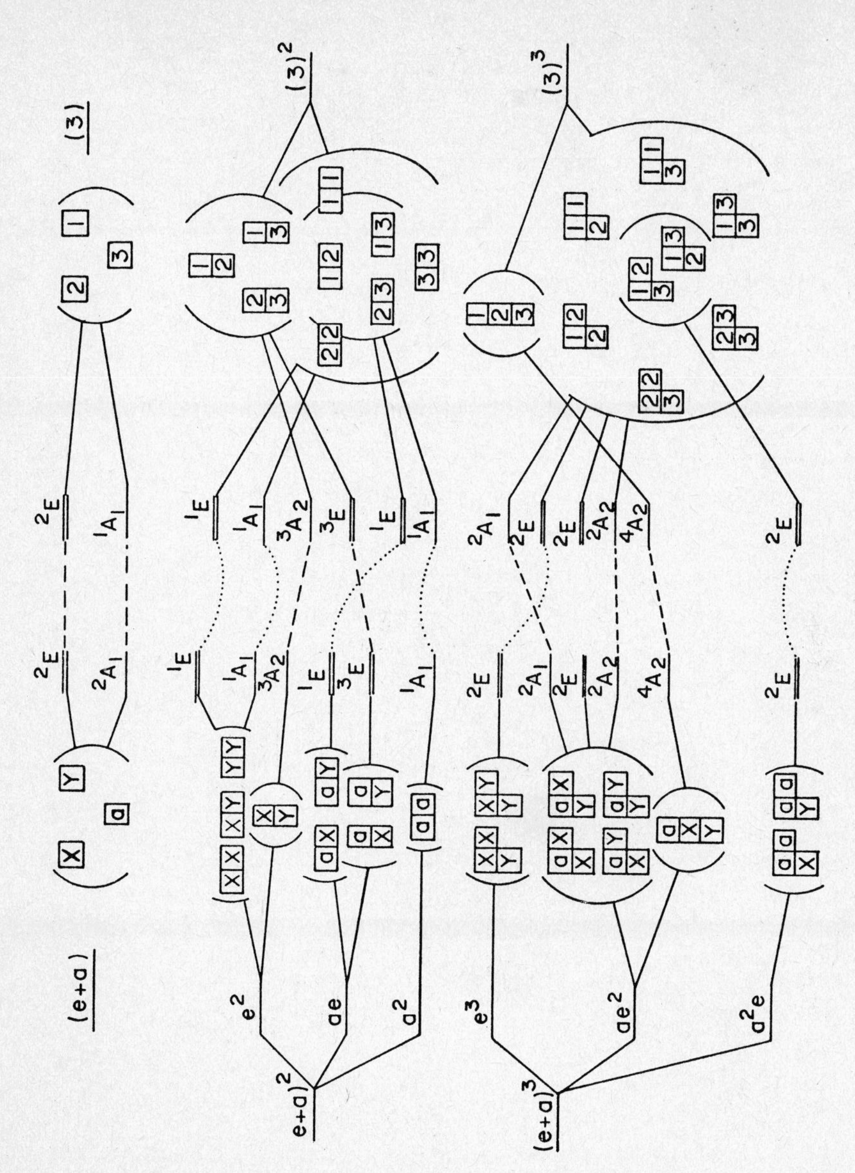

Fig.12 <u>Example of Unitary Tableau Notation for Molecular Electronic States</u>.
An orbital basis of definite spin for n electrons (n=1,2,and 3) around three fixed nuclei are indicated by tableaus. On the right the numbers refer to three equivalent valence orbitals around nucleus 1,2 or 3. At the left the states are made from products of three molecular orbitals a,x, or y. The calculation of orbital energy matrices can be done using the graphs as explained in the text.

Fig.13 <u>Associated Gelfand Basis</u>

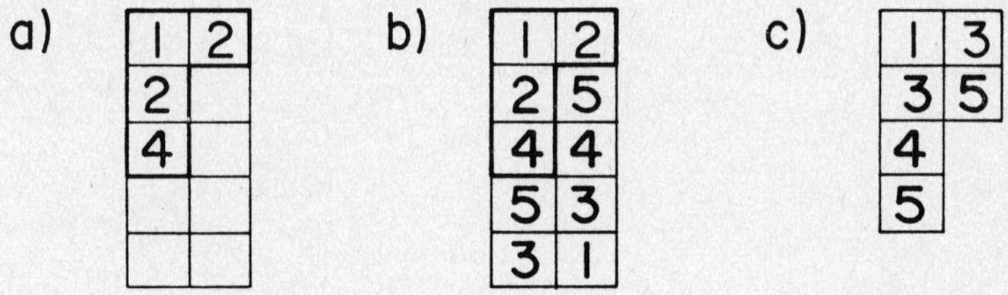

a)

1	2
2	
4	

b)

1	2
2	5
4	4
5	3
3	1

c)

1	3
3	5
4	
5	

a) $|a^*\rangle$ is found by completing a rectangle about $|a\rangle$ with 2 columns and $2\ell+1$ rows.

b) The empty boxes are numbered increasing upward in the columns with no number in a column repeated.

c) $|a\rangle$ is then detached from the rectangle and the remaining pattern rotated to give $|a^*\rangle$.

Fig.14 Matrix Elements for Associated Bases.

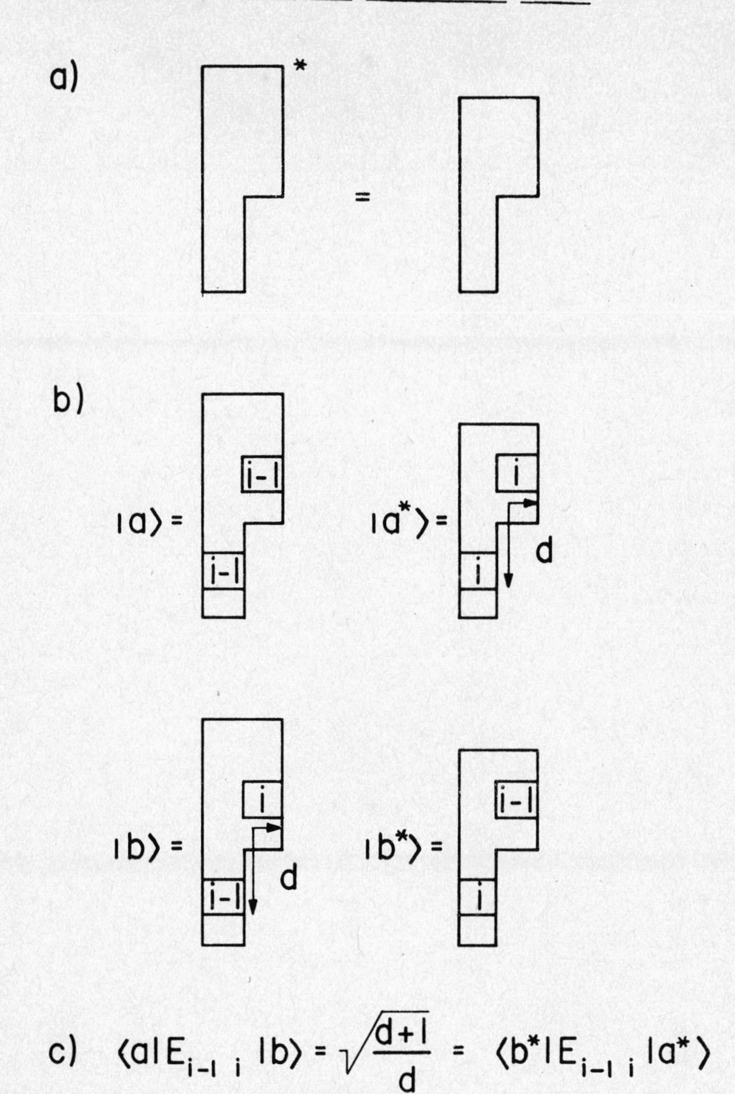

$$\text{c)} \quad \langle a | E_{i-1\,i} | b \rangle = \sqrt{\frac{d+1}{d}} = \langle b^* | E_{i-1\,i} | a^* \rangle$$

a) Because of the prescriptions a) to c) the length of both columns of a tableau and its associated tableau must differ by the same number of boxes.

b) Application of prescriptions a) to c) to states $|a\rangle$ and $|b\rangle$ to find their associated states $|a^*\rangle$ and $|b^*\rangle$. Note that the axial distance d must be preserved.

c) Equation c) in Fig. 10 is shown to verify Eq. C-2

Fig.15 Associated Basis for Antisymmetric and Spin States

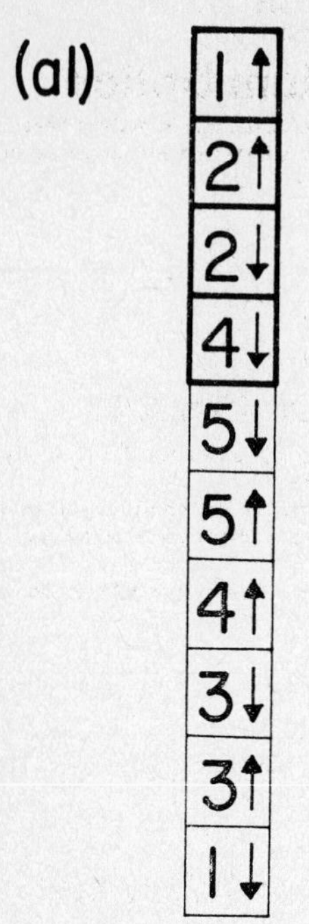

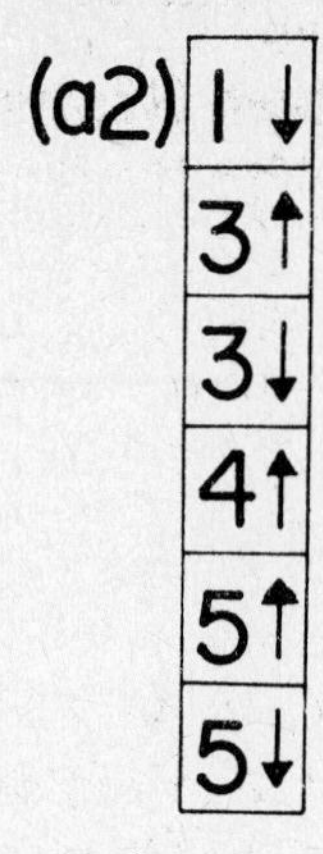

a1) A rectangle is completed about the antisymmetric Gelfand state of SU(4ℓ+2) with 1 column and 4ℓ+2 rows. The empty boxes are numbered lexically as shown. Note that n↑ > n↓.

a2) The state is then detached from the rectangle and the remaining pattern rotated to give the associated state.

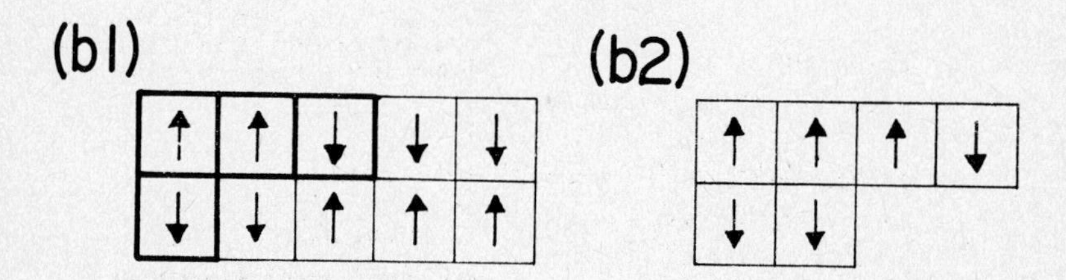

b1) A rectangle is completed about the Gelfand state of SU(2) with 2ℓ+1 columns and 2 rows. The empty boxes are numbered lexically as shown. Note that ↑ > ↓.

b2) The state is then detached from the rectangle and the remaining pattern rotated to give the associated state.

SPRINGER TRACTS IN MODERN PHYSICS

Ergebnisse der exakten Naturwissenschaften

Springer-Verlag
Berlin
Heidelberg
New York

Volume 66

30 figures. III, 173 pages. 1973
ISBN 3-540-06189-4

Quantum Statistics

in Optics and Solid-State Physics

R. Graham: Statistical Theory of Instabilities in Stationary Nonequilibrium Systems with Applications to Lasers and Nonlinear Optics.
F. Haake: Statistical Treatment of Open Systems by Generalized Master Equations.

Volume 67

III, 69 pages. 1973
ISBN 3-540-06216-5

S. Ferrara, R. Gatto, A. F. Grillo:

Conformal Algebra in Space-Time

and Operator Product Expansion

Introduction to the Conformal Group in Space-Time. Broken Conformal Symmetry. Restrictions from Conformal Covariance on Equal-Time Commutators. Manifestly Conformal Covariant Structure of Space-Time. Conformal Invariant Vacuum Expectation Values. Operator Products and Conformal Invariance on the Light-Cone. Consequences of Exact Conformal Symmetry on Operator Product Expansions. Conclusions and Outlook.

Volume 68

77 figures. 48 tables. III, 205 pages. 1973
ISBN 3-540-06341-2

Solid-State Physics

D. Schmid: Nuclear Magnetic Double Resonance — Principles and Applications in Solid-State Physics.
D. Bäuerle: Vibrational Spectra of Electron and Hydrogen Centers in Ionic Crystals.
J. Behringer: Factor Group Analysis Revisited and Unified.

Volume 69

13 figures. III, 121 pages. 1973
ISBN 3-540-06376-5

Astrophysics

G. Börner: On the Properties of Matter in Neutron Stars.
J. Stewart, M. Walker: Black Holes: the Outside Story.

Volume 70

II, 135 pages. 1974
ISBN 3-540-06630-6

Quantum Optics

G. S. Agarwal: Quantum Statistical Theories of Spontaneous Emission and their Relation to Other Approaches.

Volume 71

116 figures. III, 245 pages. 1974
ISBN 3-540-06641-1

Nuclear Physics

H. Überall: Study of Nuclear Structure by Muon Capture.
P. Singer: Emission of Particles Following Muon Capture in Intermediate and Heavy Nuclei.
J. S. Levinger: The Two and Three Body Problem.

Volume 72

32 figures. II, 145 pages. 1974
ISBN 3-540-06742-6

D. Langbein:

Theory of Van der Waals Attraction

Introduction. Pair Interactions. Multiplet Interactions. Macroscopic Particles. Retardation. Retarded Dispersion Energy. Schrödinger Formalism. Electrons and Photons.

Volume 73

110 figures. VI, 303 pages. 1975
ISBN 3-540-06943-7

Excitons at High Density

Editors: H. Haken, S. Nikitine
Biexcitons. Electron-Hole Droplets. Biexcitons and Droplets. Special Optical Properties of Excitons at High Density. Laser Action of Excitons. Excitonic Polaritons at Higher Densities.

Volume 74

75 figures. III, 153 pages. 1974
ISBN 3-540-06946-1

Solid-State Physics

G. Bauer: Determination of Electron Temperatures and of Hot Electron Distribution Functions in Semiconductors.
G. Borstel, H. J. Falge, A. Otto: Surface and Bulk Phonon-Polaritons Observed by Attenuated Total Reflection.

Selected Issues from

Lecture Notes in Mathematics

Vol. 507: M. C. Reed, Abstract Non-Linear Wave Equations. VI, 128 pages. 1976.

Vol. 501: Spline Functions, Karlsruhe 1975. Proceedings. Edited by K. Böhmer, G. Meinardus, and W. Schempp. VI, 421 pages. 1976.

Vol. 495: A. Kerber, Representations of Permutation Groups II. V, 175 pages. 1975.

Vol. 490: The Geometry of Metric and Linear Spaces. Proceedings 1974. Edited by L. M. Kelly. X, 244 pages. 1975.

Vol. 489: J. Bair and R. Fourneau, Etude Géométrique des Espaces Vectoriels. Une Introduction. VII, 185 pages. 1975.

Vol. 485: J. Diestel, Geometry of Banach Spaces – Selected Topics. XI, 282 pages. 1975.

Vol. 484: Differential Topology and Geometry. Proceedings 1974. Edited by G. P. Joubert, R. P. Moussu, and R. H. Roussarie. IX, 287 pages. 1975.

Vol. 481: M. de Guzmán, Differentiation of Integrals in R^n. XII, 226 pages. 1975.

Vol. 480: X. M. Fernique, J. P. Conze et J. Gani, Ecole d'Eté de Probabilités de Saint-Flour IV–1974. Edité par P.-L. Hennequin. XI, 293 pages. 1975.

Vol. 477: Optimization and Optimal Control. Proceedings 1974. Edited by R. Bulirsch, W. Oettli, and J. Stoer. VII, 294 pages. 1975.

Vol. 474: Séminaire Pierre Lelong (Analyse) Année 1973/74. Edité par P. Lelong. VI, 182 pages. 1975.

Vol. 470: R. Bowen, Equilibrium States and the Ergodic Theory of Anosov Diffeomorphisms. III, 108 pages. 1975.

Vol. 468: Dynamical Systems – Warwick 1974. Proceedings 1973/74. Edited by A. Manning. X, 405 pages. 1975.

Vol. 464: C. Rockland, Hypoellipticity and Eigenvalue Asymptotics. III, 171 pages. 1975.

Vol. 463: H.-H. Kuo, Gaussian Measures in Banach Spaces. VI, 224 pages. 1975.

Vol. 461: Computational Mechanics. Proceedings 1974. Edited by J. T. Oden. VII, 328 pages. 1975.

Vol. 459: Fourier Integral Operators and Partial Differential Equations. Proceedings 1974. Edited by J. Chazarain. VI, 372 pages. 1975.

Vol. 458: P. Walters, Ergodic Theory – Introductory Lectures. VI, 198 pages. 1975.

Vol. 449: Hyperfunctions and Theoretical Physics. Proceedings 1973. Edited by F. Pham. IV, 218 pages. 1975.

Vol. 448: Spectral Theory and Differential Equations. Proceedings 1974. Edited by W. N. Everitt. XII, 321 pages. 1975.

Vol. 447: S. Toledo, Tableau Systems for First Order Number Theory and Certain Higher Order Theories. III, 339 pages. 1975.

Vol. 446: Partial Differential Equations and Related Topics. Proceedings 1974. Edited by J. A. Goldstein. IV, 389 pages. 1975.

Vol. 445: Model Theory and Topoi. Edited by F. W. Lawvere, C. Maurer, and G. C. Wraith. III, 354 pages. 1975.

Vol. 444: F. van Oystaeyen, Prime Spectra in Non-Commutative Algebra. V, 128 pages. 1975.

Vol. 443: M. Lazard, Commutative Formal Groups. II, 236 pages. 1975.

Vol. 442: C. H. Wilcox, Scattering Theory for the d'Alembert Equation in Exterior Domains. III, 184 pages. 1975.

Vol. 441: N. Jacobson, PI-Algebras. An Introduction. V, 115 pages. 1975.

Vol. 440: R. K. Getoor, Markov Processes: Ray Processes and Right Processes. V, 118 pages. 1975.

Vol. 439: K. Ueno, Classification Theory of Algebraic Varieties and Compact Complex Spaces. XIX, 278 pages. 1975.

Vol. 438: Geometric Topology. Proceedings 1974. Edited by L. C. Glaser and T. B. Rushing. X, 459 pages. 1975.

Vol. 437: D. W. Masser, Elliptic Functions and Transcendence. XIV, 143 pages. 1975.

Vol. 436: L. Auslander and R. Tolimieri, Abelian Harmonic Analysis, Theta Functions and Function Algebras on a Nilmanifold. V, 99 pages. 1975.

Vol. 435: C. F. Dunkl and D. E. Ramirez, Representations of Commutative Semitopological Semigroups. VI, 181 pages. 1975.

Vol. 434: P. Brenner, V. Thomée, and L. B. Wahlbin, Besov Spaces and Applications to Difference Methods for Initial Value Problems. II, 154 pages. 1975.

Vol. 433: W. G. Faris, Self-Adjoint Operators. VII, 115 pages. 1975.

Vol. 432: R. P. Pflug, Holomorphiegebiete, pseudokonvexe Gebiete und das Levi-Problem. VI, 210 Seiten. 1975.

Vol. 431: Séminaire Bourbaki – vol. 1973/74. Exposés 436–452. IV, 347 pages. 1975.

Vol. 430: Constructive and Computational Methods for Differential and Integral Equations. Proceedings 1974. Edited by D. L. Colton and R. P. Gilbert. VII, 476 pages. 1974.

Vol. 429: L. Cohn, Analytic Theory of the Harish-Chandra C-Function. III, 154 pages. 1974.

Vol. 428: Algebraic and Geometrical Methods in Topology, Proceedings 1973. Edited by L. F. McAuley. XI, 280 pages. 1974.

Vol. 427: H. Omori, Infinite Dimensional Lie Transformation Groups. XII, 149 pages. 1974.

Vol. 426: M. L. Silverstein, Symmetric Markov Processes. X, 287 pages. 1974.